RANKING TASK
EXERCISES IN PHYSICS:
Student Edition

edited by

Thomas L. O'Kuma
Lee College
Baytown, Texas

David P. Maloney
Indiana University-Purdue University Fort Wayne
Fort Wayne, Indiana

Curtis J. Hieggelke
Joliet Junior College
Joliet, Illinois

PEARSON
Addison
Wesley

Boston   San Francisco   New York
London   Toronto   Sydney   Tokyo   Singapore   Madrid
Mexico City   Munich   Paris   Cape Town   Hong Kong   Montreal

The Library of Congress has cataloged a previous edition as follows:

Ranking task exercises in physics: student edition / edited by Thomas L. O'Kuma, David P. Maloney, Curtis J. Hieggelke.

    p. cm .- - (Prentice Hall series in educational innovation)
    Includes bibliographical references
    ISBN 0-13-022355-7
    1. Physics Problems, exercises, etc. I. O'Kuma, Thomas L, II.
Maloney, David P.   III. Hieggelke, Curtis J.   IV. Series.
    QC32 .R28 1999
    530' 076 dc21, 99-32915
,CIP

Senior Editor: Erik Fahlgren
Editor in Chief, Science: John Challice
Vice President of Production and Manufacturing: David W. Riccardi
Executive Managing Editor: Kathleen Schiaparelli
Production Editor: Edward Thomas
Manufacturing Manager: Trudy Pisciotti
Manufacturing Buyer: Alan Fischer
Executive Marketing Manager: Mark Pfaltzgraff
Editorial Assistant: Andrew Sobel
Proofreader: Karen Bosch
Production Assistant: Nancy Bauer
Art Director: Jayne Conte
Cover Designer: Lisa Boylan

10 9 8 7 6 5

ISBN 0-13-144851-X

Pearson Education Ltd., London
Pearson Education Australia Pty., Limited, Sydney
Pearson Education Singapore, Pte. Ltd.
Pearson Education North Asia Ltd., Hong Kong
Pearson Education Canada, Ltd., Toronto
Pearson Educación de Mexico, S.A. de C.V.
Pearson Education—Japan, Tokyo
Pearson Education Malaysia, Pte. Ltd.

About the Authors

Since 1991, Thomas O'Kuma, David Maloney, and Curtis Hieggelke have led over 20 workshops in which educators learned how to use and develop ranking tasks. Many of these workshops were part of the Two Year College Physics Workshop Project (supported by seven grants from the National Science Foundation and directed by Curtis Hieggelke and Thomas O'Kuma), which has offered a series of more than 50 professional development workshops for over 1000 Two-Year College and High School physics teacher participants. Working with Alan Van Heuvelen, they also developed the Conceptual Survey of Electricity and Magnetism, which has become a standard instrument for measuring electricity and magnetism conceptual gains in introductory physics courses.

Thomas L. O'Kuma received his B.S. and M.S. in physics and mathematics from Louisiana Tech University. Since 1976 he has been a full-time, two-year college physics instructor, and has been at Lee College since 1989. He wrote the *Study Guide to Beiser's Physics* and has written various lab manuals for the introductory physics course. He received the Distinguished Service Citation in 1994 and the Award for Excellence in Introductory College Physics Teaching in 2002 from the American Association of Physics Teachers (AAPT), and the Robert N. Little Award in 1994 for Outstanding Contributions to Physics in Higher Education in Texas. O'Kuma's research interests are in how students learn physics and in developing tools to help students learn physics.

David P. Maloney received a B.S. in physics from the University of Louisville, and a M.S. in physics and Ph.D. in physics, geology and education from Ohio University. A member of the Indiana University-Purdue University-Fort Wayne faculty since 1987, he has also taught at Wesleyan College and Creighton University. His main research interests concern the study of students' common-sense ideas about physics, how those ideas interact with physics instruction, and the study of problem-solving in physics. Maloney has authored or co-authored two-dozen articles and has been the Principal Investigator or Co-Principal Investigator for eight National Science Foundation (NSF) grant projects.

Curtis J. Hieggelke received a B.A. in physics and mathematics from Concordia College (Minnesota) and a Ph.D in theoretical particle physics from the University of Nebraska. He spent his entire professional career as a full-time physics teacher at Joliet Junior College. Hieggelke has served as President and Section Representative of the Illinois Section of the AAPT (ISAAPT), and received the Distinguished Service Citation in 1993. Joliet Junior College received the Illinois Community College Board Excellence in Teaching/Learning Award for his work in 1992. He was awarded the Distinguished Service Citation by AAPT in 1994 and he was elected to the Executive Board of AAPT for three years as the Two-Year College Representative. He has received three NSF grants for laboratory curriculum development. His current research work in electrostatics and magnetism is supported by two NSF grants for curriculum materials development.

Contents

SHM and Properties of Matter Ranking Tasks

Heat and Thermodynamics Ranking Tasks

Electrostatics Ranking Tasks

Wave and Optics Ranking Tasks

Preface

This workbook is designed to be a resource for students to deepen their understanding of physics. Students need to get beyond the "memorize and recall" approach and this workbook will be useful for that purpose. The tasks in this manual ask the user to apply the ideas that are being learned in a new way. Using the ideas in different ways will get the user to think about those ideas in more breadth and depth, leading to a stronger understanding.

The tasks in this manual grew out of investigations into students thinking about physics as they were learning the subject. These tasks are likely to be quite different from the tasks you have encountered in other math and science courses. They are different in several ways. First, comparing several similar situations to each other, rather than finding a particular value associated with a situation, is likely to be new. This novelty may make doing ranking tasks uncomfortable since you may feel insecure about how to determine such comparisons. Practice will make this reaction fade.

A second way these tasks differ from more familiar ones is that a number of these tasks will contain irrelevant information. Since you are probably used to problems where all information provided needs to be used, the job of having to determine whether, and what, information may be irrelevant may be challenging until you get used to it. But you should be aware that problems in real life very often require determining what information is needed and what information can be ignored. Since that is the case it is worthwhile to have some practice with such problems in academic settings also.

Another difference is that these tasks ask you to determine and write a ranking, then explain the reasoning/work behind your ranking sequence, and finally indicate a confidence level for your work. It is important to explicitly write out your explanations, even when you are doing the items for yourself, because you will often find that your thinking will change and improve when you try to explain what you did and why you did it.

A fourth difference is that there are groups of related tasks throughout the workbook. The tasks in these groupings can be related in a number of ways. For example, the situations in two tasks may be exactly the same, but the basis for the ranking may be different. Thinking about the similarities and differences among tasks in such groupings can provide very useful insights into the physics of the situations.

This workbook is designed to be useful for students in all introductory-level physics courses, so some of the tasks included may involve ideas, and/or mathematics, specific to other courses. Please do not be concerned if some of the tasks seem not related to your course or background.

Since the format for the ranking tasks in this workbook is most likely unfamiliar to you, you would be well advised to spend a little time getting used to the format itself before trying to deal with these tasks. To aid in the familiarization process we have provided several example tasks showing how to complete the tasks correctly. We do strongly encourage you to read and study these introductory materials so that you feel comfortable with the structure and response process for ranking tasks before you start these physics ranking tasks.

When you approach a ranking task you should not try to think of some equation that uses the quantities given and then plug the values in to find the "answers." As mentioned above, many of the ranking tasks contain irrelevant information, in some cases all of the numerical values provided are irrelevant. So if you automatically take a "plug and chug" approach to the ranking tasks you will find that you are often going to be doing calculations for nothing. The best way to tackle a ranking task is to think about how you expect the object/system in the physical situation to behave. Then use the given information to determine the specific comparisons among the given variations.

Acknowledgments

First, we want to thank the National Science Foundation for their support of the Two-Year College Physics Workshop Projects. Also appreciated is the strong support by Joliet Junior College and Lee College. We want to acknowledge and thank all the participants who attended our workshops and were willing to share their ideas and work along with the workshop leaders. We also want to thank our students, who tried out these ranking tasks and provided us with valuable feedback and insights. Suggestions from William Hogan (Joliet Junior College, Joliet, IL), and Duane Desbien (Estrella Mountain Community College, Avondale, AZ) helped improve these ranking tasks.

We are particularly grateful for the ideas and ranking tasks from Dennis Albers (Columbia College, Sonora, CA), Toni Bennie-George (Green River Community College, Auburn, WA), Light Bryant (South Mountain Community College, Phoenix, AZ), Jonathan Cole (Mira Costa Community College, Oceanside, CA), Paul D'Alessandris (Monroe Community College, Rochester, NY), Chad Davies (Cloud County Community College, Concordia, KS), Alex Dickison (Seminole Community College, Sanford, FL), Butch Diesslin (Vermilion Community College, Ely, MN), Eldon Eckard (Bainbridge College, Bainbridge, GA), Bruce Emerson (Central Oregon Community College, Bend, OR), Cathy Ezrailson (Montgomery College, Conroe, TX), Chuen Chuen Fang (Shelton State Community College, Tuscaloosa, AL), Greg Fazzari (Walla Walla Community College, Walla Walla, WA), Peter Golden (Isothermal Community College, Spindale, NC), Jim Gundlach (John A. Logan College, Carterville, IL), Sally Heath (Santa Rosa Community College, Santa Rosa, CA), Jack Hehn (Tarrant County Junior College, Hurst, TX), Charlotte Hill (Tarrant County Junior College, Hurst, TX), Robert Johnson (Del Mar College, Corpus Christi, TX), Bruce Kaasa (Iowa Central Community College, Fort Dodge, IA), Oshri Karmon (Diablo Valley College, Pleasant Hill, CA), Beta Keramati (Holmes Community College, Goodman, MS), Robert Krupp (Wilber Wright College, Chicago, IL), Clement Lam (North Harris College, Houston, TX), Stuart Loucks (Solano Community College, Suisun, CA), Myron Mann (Los Angeles Valley College, Van Nuys, CA), Paul Marquard (Casper College, Casper, WY), Marvin Nelson (Green River Community College, Auburn, WA), Nick Nicholson (Central Alabama Community College, Alexander City, AL), Umesh Pandey (Albuquerque Technical–Vocational Institute, Albuquerque, NM), Marie Plumb (Jamestown Community College, Jamestown, NY), John Popp (Moraine Valley Community College, Palos Hills, IL), Sherry Saavrda (Lake Sumter Community College, Leesburg, FL), Dave Schramme (Rogers State College, Claremore, OK), Ot Selgrad (Moraine Park Technical College, Fond du Lac, WI), Gordon Shepherd (Guilford Technical Community College, Jamestown, NC), Bob Speers (Fireland College – Bowling Green St. University, Huron, OH), John Splett (Erie Community College – City Campus, Buffalo, NY), Imre Szalai (Delaware County Community College, Media, PA), Leo Takahashi (Penn State – Beaver Campus, Monaca, PA), Jack Taylor (Baltimore City Community College, Baltimore, MD), Winston Thompson (Montgomery College, Takoma Park, MD), Cheng (David) Ting (Southeast College–Houston Community College, Houston, TX), Alan Van Heuvelen (The Ohio State University, Columbus, OH), Adam Weiner (Green River Community College, Auburn, WA), Myra West (Kent State University, Kent, OH), Karen Wetz (Manatee Community College, Venice, FL), and Bob Willis (New Mexico Military Institute, Roswell, NM.)

We thank Steve Kanim (New Mexico State University, Las Cruces, NM) for his superb drawing for this edition. Finally, at Prentice Hall, we appreciate those who supported this project and the publication of this book. This includes Erik Fahlgren, Christian Botting, and Andrew Sobel.

This material is based upon work supported by the Division of Undergraduate Education of the National Science Foundation under grant numbers USE 9150334, USE 9154271, DUE 9255466, DUE 9353998, DUE 9554683, DUE 9952735, and DUE 0125831. Any opinions, findings, and conclusions or recommendations are those of the authors and do not necessarily reflect the views of the National Science Foundation.

Ranking Task Sample I

For a ranking task, each item will have a number of situations as illustrated. Your task will be to rank the items in a specific order. After ranking them you will be asked to identify the basis you used for the ranking and the reasoning behind your choice. It is extremely important that you are careful to write out the proper ranking once you have determined what basis you are going to use, i.e., make sure all of the situations are ranked in the proper order according to your basis. The sample below shows how to rank items and what your explanation should be like. **NOTE: Although the procedure for working the item is correct, the particular answer, which was chosen at random from actual student responses, may not be correct.**

Example:

Shown below are eight cars that are moving along horizontal roads at specified speeds. Also given are the masses of the cars. All of the cars are the same size and shape, but they are carrying loads with different masses. All of these cars are going to be stopped by plowing into barrel barriers. All of the cars are going to be stopped in the same distance.

Rank these situations from greatest to least on the basis of the strength of the forces that will be needed to stop the cars in the same distance. That is, put first the car on which the strongest force will have to be applied to stop it in x meters, and put last the car on which the weakest force will be applied to stop the car in the same distance.

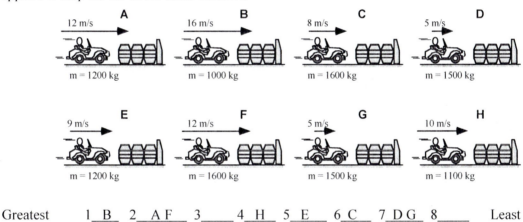

Greatest   1 _B_   2 _A F_   3____   4 _H_   5 _E_   6 _C_   7 _D G_   8____   Least

Or, all cars require the same force. _____

Please carefully explain your reasoning.

Since acceleration is the change in velocity divided by the change in time and all the changes in times are the same, then I used the change in velocity.

How sure were you of your ranking? (circle one)

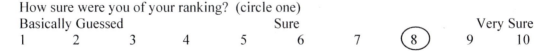

Notice in this example that two situations produced the same result for the ranking and that these were listed in the same answer blank. Such a possibility exists for all items. In the same way, it is possible that all of the situations will give the same result. If that occurs, and only if that occurs, the option of all equal, or all the same, should be chosen.

Ranking Task Sample II

Each ranking task will have a number of situations, or variations of a situation, that have varying values for two or three variables. Your task is to rank these variations on a specified basis. After ranking the items, you will be asked to explain how you determined your ranking sequence and the reasoning behind the way you used the values of the variables to reach your answer. An example of how to work the ranking tasks follows.

Example:

Shown below are six situations where a cart, which is initially moving to the right, has a force applied to it such that the force will cause the cart to come to a stop. All of the carts have the same initial speed, but the masses of the carts vary, as do the forces acting upon them.

Rank these situations, from greatest to least, on the basis of how long it will take each cart to stop.

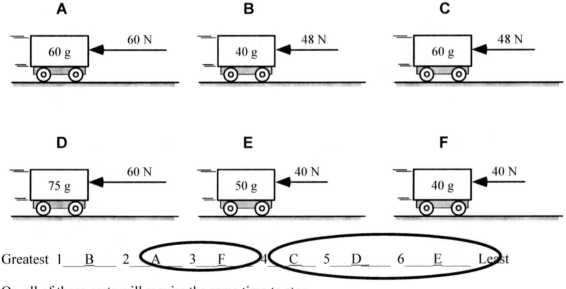

Greatest 1 __B__ 2 __A__ 3 __F__ 4 __C__ 5 __D__ 6 __E__ Least

Or, all of these carts will require the same time to stop. _____

Please carefully explain your reasoning.

I think the time depends on the acceleration, so I divided the forces by the masses.

How sure were you of your ranking? (circle one)
Basically Guessed   Sure   Very Sure
1   2   3   4   5   6   7   8   9   10

Notice in this example that in one instance, two of the situations produced the same value of the ratio used to determine the ranking, and that the letters for the ones that tied are circled showing they were ranked equally (A and F). In another instance, three of the remaining situations have the same ranking and they are circled together (C and D and E), showing this result. In the same way, it is possible that all of the arrangements will give the same result for a particular basis. If that occurs, and only if that occurs, the option of all equal, or all the same, should be chosen.

Ball Motion Diagrams—Velocity I[1]

The following drawings indicate the motion of a ball subject to one or more forces on various surfaces from <u>left to right</u>. Each circle represents the position of the ball at succeeding instants of time. Each time-interval between successive positions is equal.

Rank each case from the highest to the lowest velocity based on the ball's last velocity using the coordinate system specified in the figures. Note: Zero is greater than negative, and ties are possible.

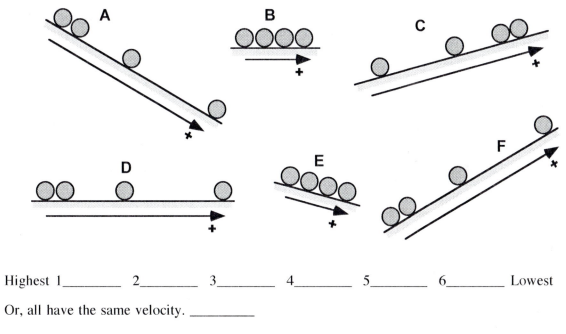

Highest 1_____ 2_____ 3_____ 4_____ 5_____ 6_____ Lowest

Or, all have the same velocity. _____

Please carefully explain your reasoning.

How sure were you of your ranking? (circle one)

Basically Guessed				Sure				Very Sure	
1	2	3	4	5	6	7	8	9	10

[1] D. Schramme, C. Fang, B. Speers
Ranking Task Exercises in Physics: Student Edition   1

Ball Motion Diagrams—Acceleration I[2]

The following drawings indicate the motion of a ball subject to one or more forces on various surfaces from <u>left to right</u>. Each circle represents the position of the ball at succeeding instants of time. Each time-interval between successive positions is equal.

Rank each case from the highest to the lowest acceleration, based on the drawings. Assume all accelerations are constant and use the coordinate system specified in the drawing. Note: Zero is greater than negative acceleration, and ties are possible.

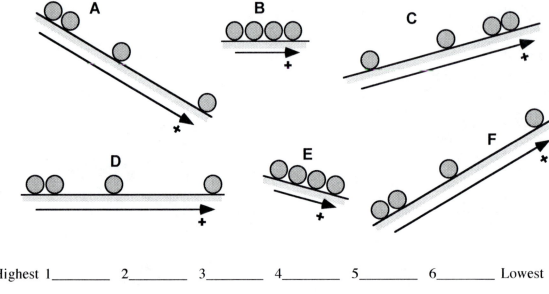

Highest 1_____ 2_____ 3_____ 4_____ 5_____ 6_____ Lowest

Or, all have the same acceleration. _____

Please carefully explain your reasoning.

How sure were you of your ranking? (circle one)

Basically Guessed					Sure				Very Sure	
1	2	3	4	5	6	7	8	9	10

[2] D. Schramme, C. Fang, B. Speers
Ranking Task Exercises in Physics: Student Edition   2

Ball Motion Diagrams—Velocity II[3]

The following drawings indicate the motion of a ball subject to one or more forces on various surfaces from <u>left to right</u>. Each circle represents the position of the ball at succeeding instants of time. Each time-interval between successive positions is equal.

Rank each case from the highest to the lowest velocity based on the ball's last velocity using the coordinate system specified in the drawing. Note: Zero is greater than negative, and ties are possible.

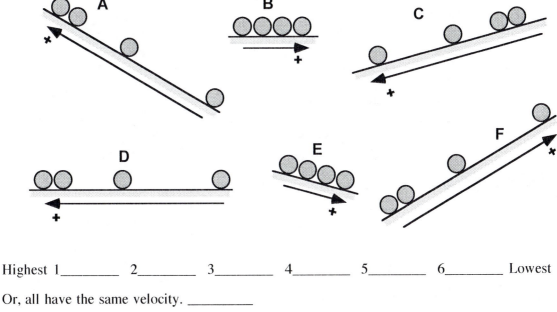

Highest 1_____ 2_____ 3_____ 4_____ 5_____ 6_____ Lowest

Or, all have the same velocity. _____

Please carefully explain your reasoning.

How sure were you of your ranking? (circle one)

Basically Guessed				Sure				Very Sure	
1	2	3	4	5	6	7	8	9	10

[3] D. Schramme, C. Fang, B. Speers, C. Hieggelke, D. Maloney, T. O'Kuma
Ranking Task Exercises in Physics: Student Edition   3

Ball Motion Diagrams—Acceleration II[4]

The following drawings indicate the motion of a ball subject to one or more forces on various surfaces from <u>left to right</u>. Each circle represents the position of the ball at succeeding instants of time. Each time-interval between successive positions is equal.

Rank each case from the highest to the lowest acceleration, based on the drawings. Assume all accelerations are constant and use the coordinate system specified in the drawing. Note: Zero is greater than negative acceleration, and ties are possible.

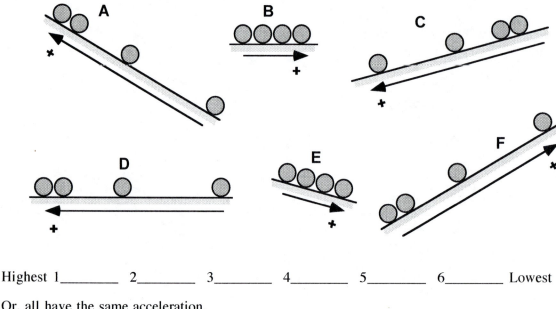

Highest 1_____ 2_____ 3_____ 4_____ 5_____ 6_____ Lowest

Or, all have the same acceleration. _____

Please carefully explain your reasoning.

How sure were you of your ranking? (circle one)
Basically Guessed   Sure   Very Sure
1   2   3   4   5   6   7   8   9   10

[4] D. Schramme, C. Fang, B. Speers, C. Hieggelke, D. Maloney, T. O'Kuma
Ranking Task Exercises in Physics: Student Edition   4

Objects in Different Situations—Accelerations[5]

The following objects are sitting, falling, rolling, swinging, or just going in circles, as indicated in the different situations below. Each may have an acceleration in some direction or another.

Rank the different situations from greatest to least on the basis of the magnitude of acceleration. If two or more objects have the same acceleration, rank them the same.

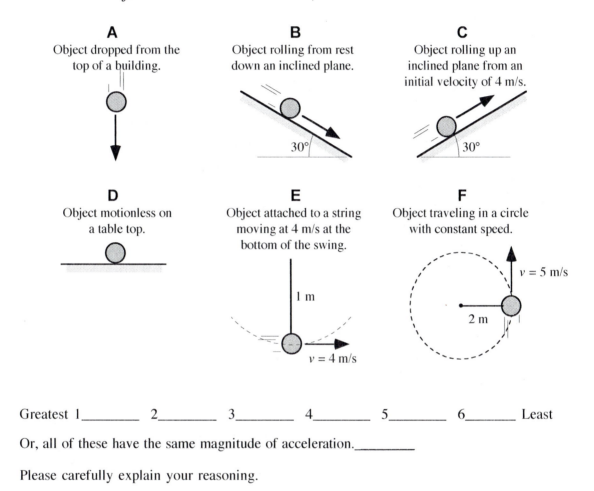

A
Object dropped from the top of a building.

B
Object rolling from rest down an inclined plane.

C
Object rolling up an inclined plane from an initial velocity of 4 m/s.

D
Object motionless on a table top.

E
Object attached to a string moving at 4 m/s at the bottom of the swing.

F
Object traveling in a circle with constant speed.

Greatest 1_____ 2_____ 3_____ 4_____ 5_____ 6_____ Least

Or, all of these have the same magnitude of acceleration._____

Please carefully explain your reasoning.

How sure were you of your ranking? (circle one)

Basically Guessed				Sure				Very Sure	
1	2	3	4	5	6	7	8	9	10

Vertical Model Rockets—Maximum Height[6]

The six figures below depict six model rockets that have just had their engines turned off. All of the rockets are aimed straight up, but their speeds differ. All of the rockets are the same size and shape, but they carry different loads, so their masses differ. The specific mass and speed for each rocket is given in each figure. (In this situation, we are going to ignore any effect air resistance may have on the rockets.) At the instant when the engines are turned off, the rockets are all at the same height.

Rank these model rockets, from greatest to least, on the basis of the maximum height they will reach.

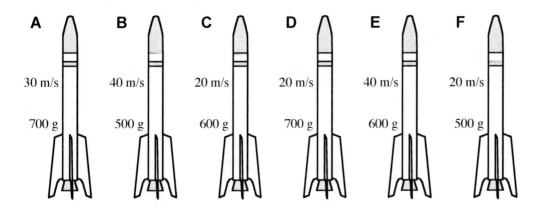

A	B	C	D	E	F
30 m/s	40 m/s	20 m/s	20 m/s	40 m/s	20 m/s
700 g	500 g	600 g	700 g	600 g	500 g

Greatest 1_____ 2_____ 3_____ 4_____ 5_____ 6_____ Least

Or, all rockets reach the same height._____

Please carefully explain your reasoning.

How sure were you of your ranking? (circle one)
Basically Guessed   Sure   Very Sure
1   2   3   4   5   6   7   8   9   10

[6] T. O'Kuma
Ranking Task Exercises in Physics: Student Edition   6

Vertical Arrows—Maximum Height[7]

The six figures below show arrows that have been shot into the air. All of the arrows were shot straight up and are the same size and shape. But the arrows are made of different materials so they have different masses, and they have different speeds as they leave the bows. The values for each arrow are given in the figures. (We assume for this situation that the effect of air resistance can be neglected.) All start from the same height.

Rank these arrows, from greatest to least, on the basis of the maximum heights the arrows reach.

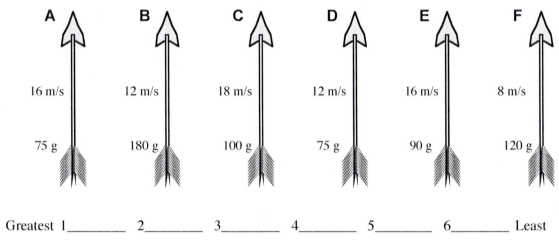

Greatest 1_____ 2_____ 3_____ 4_____ 5_____ 6_____ Least

Or, all the arrows reach the same height. _____

Please carefully explain your reasoning.

How sure were you of your ranking? (circle one)
Basically Guessed   Sure   Very Sure
1   2   3   4   5   6   7   8   9   10

Position Time Graphs—Displacement[8]

In the position vs. time graphs below, all the times are in seconds (s), and all the positions are in meters (m).

Rank these graphs on the basis of which graph indicates the greatest displacement from beginning to end of motion. Give the highest rank to the one(s) with the greatest displacement, and give the lowest rank to the one(s) indicating the least displacement. Note: Zero is greater than negative, and ties are possible.

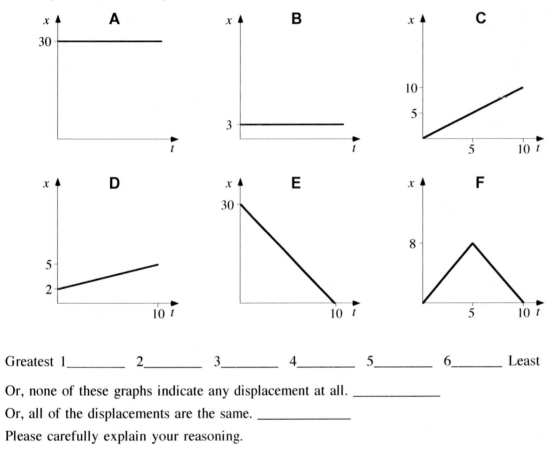

Greatest 1_____ 2_____ 3_____ 4_____ 5_____ 6_____ Least

Or, none of these graphs indicate any displacement at all. _____

Or, all of the displacements are the same. _____

Please carefully explain your reasoning.

How sure were you of your ranking? (circle one)
Basically Guessed   Sure   Very Sure
1   2   3   4   5   6   7   8   9   10

8 K. W. Nicholson

Cars—Change of Velocity[9]

The six situations below show *before* and *after* "snapshots" of a car's velocity. All cars have the same mass and they traveled the same distance.

Rank these situations, in terms of the change in velocity, from most positive to most negative. Negative numbers, if any, rank lower than positive ones (-20 m/s < -10 m/s < 0 < 5 m/s).

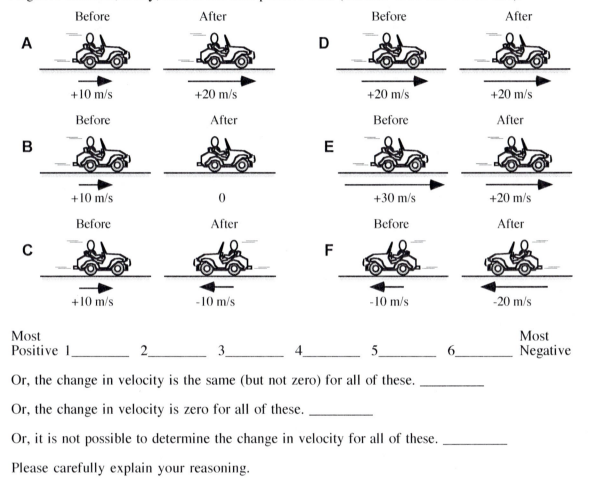

Most
Positive 1_____ 2_____ 3_____ 4_____ 5_____ 6_____ Most Negative

Or, the change in velocity is the same (but not zero) for all of these. _____

Or, the change in velocity is zero for all of these. _____

Or, it is not possible to determine the change in velocity for all of these. _____

Please carefully explain your reasoning.

How sure were you of your ranking? (circle one)

Basically Guessed   Sure   Very Sure

1   2   3   4   5   6   7   8   9   10

Position Time Graphs—Average Speed[10]

In the position vs. time graphs below, all the times are in seconds (s), and all the positions are in meters (m).

Rank these graphs on the basis of which graph indicates the greatest average speed, where the average speed is calculated from the beginning to the end of motion. Give the highest rank to the one(s) with the greatest average speed, and give the lowest rank to the one(s) indicating the least average speed.

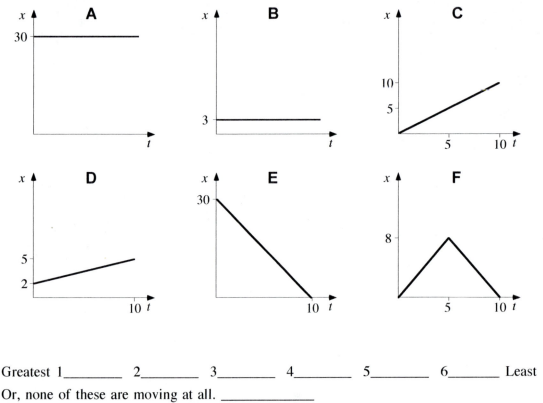

Greatest 1_____ 2_____ 3_____ 4_____ 5_____ 6_____ Least

Or, none of these are moving at all. _____

Or, the average speed is the same for all of these. _____

Please carefully explain your reasoning.

How sure were you of your ranking? (circle one)

Basically Guessed					Sure				Very Sure	
1	2	3	4	5	6	7	8	9	10

[10] K. W. Nicholson

Motion Diagrams—Displacement[11]

Flash strobe photographs were taken every second of a set of spheres moving from left to right. The diagram below shows the location of each sphere when each photograph was taken. The total time intervals shown vary among the spheres. All the displacements are in meters.

Rank these spheres on the basis of the greatest displacement over the first 3 seconds. Give the highest rank to the one(s) with the greatest displacement, and give the lowest rank to the one(s) indicating the least displacement.

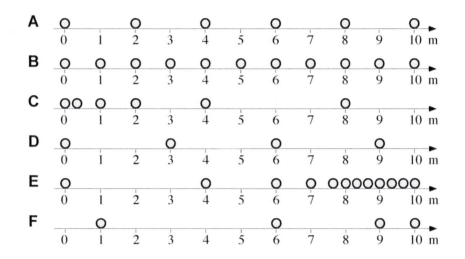

Greatest 1_____ 2_____ 3_____ 4_____ 5_____ 6_____ Least

Or, none of these motion diagrams indicate any displacement at all. _____

Or, the displacement is the same for all of these. _____

Please carefully explain your reasoning.

How sure were you of your ranking? (circle one)

Basically Guessed					Sure			Very Sure	
1	2	3	4	5	6	7	8	9	10

[11] K. W. Nicholson, C. Hieggelke, D. Maloney
Ranking Task Exercises in Physics: Student Edition   11

Motion Diagrams—Average Velocity[12]

Flash strobe photographs were taken every second of a set of spheres moving from left to right. The diagram below shows the location of each sphere when each photograph was taken. The total time intervals shown vary among the spheres. All the displacements are in meters.

Rank these spheres on the basis of the greatest average velocity over the first 3 seconds. Give the highest rank to the one(s) with the greatest average velocity, and give the lowest rank to the one(s) indicating the least average velocity.

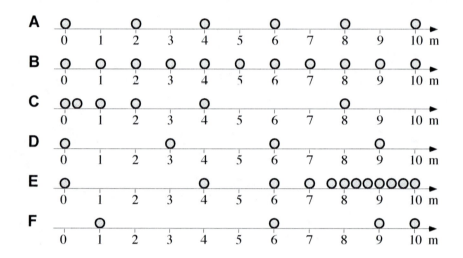

Greatest 1_____ 2_____ 3_____ 4_____ 5_____ 6_____ Least

Or, none of these are moving at all. _____

Or, the average velocity is the same for all of these. _____

Please carefully explain your reasoning.

How sure were you of your ranking? (circle one)
Basically Guessed   Sure   Very Sure
1   2   3   4   5   6   7   8   9   10

[12] K. W. Nicholson, C. Hieggelke, D. Maloney
Ranking Task Exercises in Physics: Student Edition   12

People on Trains—Speed Relative to Ground[13]

Shown below are six situations where people are walking or running along flat train cars. The people are walking/running at various speeds either in the direction the train is traveling or in the opposite direction to the train's direction. The speed of the train and the speed and direction of each person relative to the train are given in each figure. An observer is standing beside the track watching the train go by.

Rank these walkers/runners, from fastest to slowest, on the basis of how fast they are moving relative to the observer standing beside the tracks. That is, put first the person the observer would say is going fastest and put last the person the observer would say was the slowest.

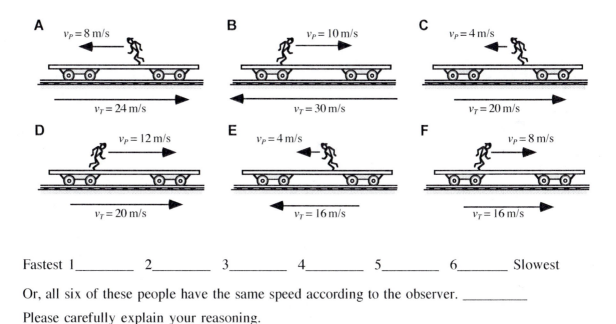

Fastest 1_____ 2_____ 3_____ 4_____ 5_____ 6_____ Slowest

Or, all six of these people have the same speed according to the observer. _____

Please carefully explain your reasoning.

How sure were you of your ranking? (circle one)

Basically Guessed				Sure				Very Sure	
1	2	3	4	5	6	7	8	9	10

[13] C. Hieggelke, D. Maloney, T. O'Kuma
Ranking Task Exercises in Physics: Student Edition   13

Carts Moving Along Horizontal Surface—String Tension[14]

The six figures below show carts that are moving along horizontal surfaces at various speeds. The carts are the same size and shape but carry different loads, so their masses differ. All of the carts have a massless string attached, which passes over a frictionless massless pulley and is tied to a metal block that is hanging free. All of the metal blocks are identical. As the carts move to the right they pull the blocks up toward the horizontal surface, which is the top of the table.

Rank these situations, from greatest to least, on the basis of the tension in the strings at the instant shown. That is, put first the situation where the string is under the greatest tension, and put last the situation where the string is under the least tension at that instant.

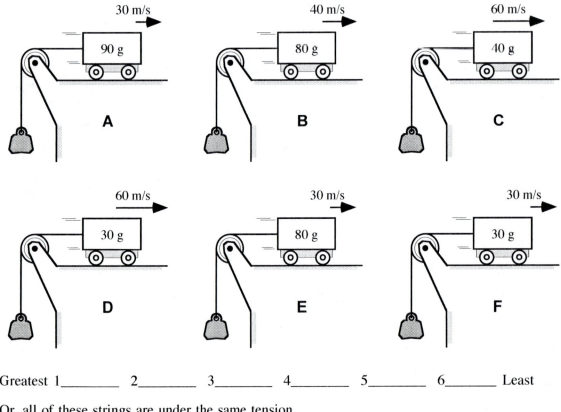

Greatest 1_____ 2_____ 3_____ 4_____ 5_____ 6_____ Least

Or, all of these strings are under the same tension. _____

Or, there is no tension in any of these strings. _____

Please carefully explain your reasoning.

How sure were you of your ranking? (circle one)
Basically Guessed   Sure   Very Sure
1   2   3   4   5   6   7   8   9   10

Carts Moving Along Horizontal Surface—Acceleration[15]

The six figures below show carts that are moving along horizontal surfaces at various speeds. The carts are the same size and shape but carry different loads, so their masses differ. All of the carts have a string attached, which passes over a pulley and is tied to a metal block that is hanging free. All of the metal blocks are identical. As the carts move to the right, they will pull the blocks up toward the horizontal surface, which is the top of the table.

Rank these situations, from greatest to least, on the basis of the magnitude of the acceleration of the carts. That is, put first the situation where the cart has the greatest acceleration, and put last the situation where the cart has the least acceleration.

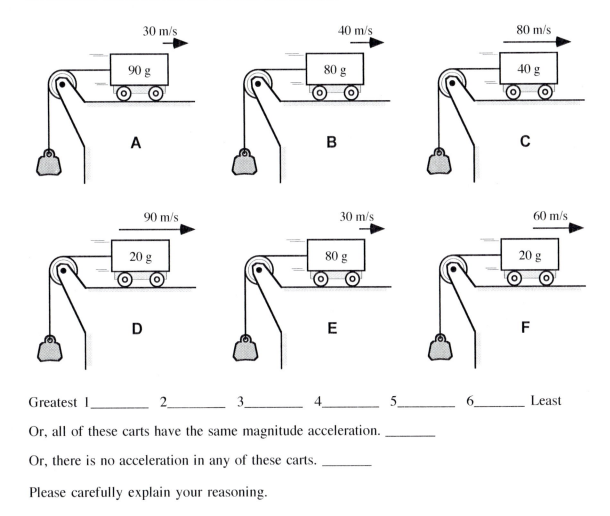

Greatest 1_____ 2_____ 3_____ 4_____ 5_____ 6_____ Least

Or, all of these carts have the same magnitude acceleration. _____

Or, there is no acceleration in any of these carts. _____

Please carefully explain your reasoning.

How sure were you of your ranking? (circle one)

Basically Guessed				Sure				Very Sure	
1	2	3	4	5	6	7	8	9	10

[15] D. Maloney

Carts Moving Along Horizontal Surface—Slowing Down[16]

The six figures below show carts that are moving along horizontal surfaces at various speeds. The carts are the same size and shape but carry different loads so their masses differ. All of the carts have a string attached, which passes over a pulley and is tied to a metal block that is hanging free. All of the metal blocks are identical. As the carts move to the right they pull the blocks up toward the horizontal surface, which is the top of the table.

Rank these situations, from greatest to least, on the basis of which cart is slowing down most quickly. That is, put first the situation where the cart is slowing down the quickest, and put last the situation where the cart is slowing down at the slowest rate.

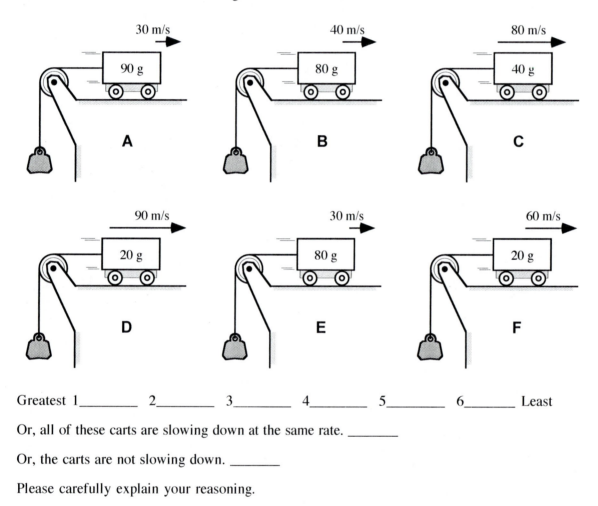

Greatest 1_____ 2_____ 3_____ 4_____ 5_____ 6_____ Least

Or, all of these carts are slowing down at the same rate. _____

Or, the carts are not slowing down. _____

Please carefully explain your reasoning.

How sure were you of your ranking? (circle one)
Basically Guessed   Sure   Very Sure
1   2   3   4   5   6   7   8   9   10

[16] D. Maloney

Two-Dimensional Forces on a Treasure Chest—Final Speed[17]

The six figures below show treasure chests with two forces acting upon them. We are looking down on the chests from above. The lengths of the force vectors represent the magnitudes of the forces.

Rank these situations from greatest to least with regard to the final speed of the treasure chest after 2 seconds. All chests start at rest and have the same mass.

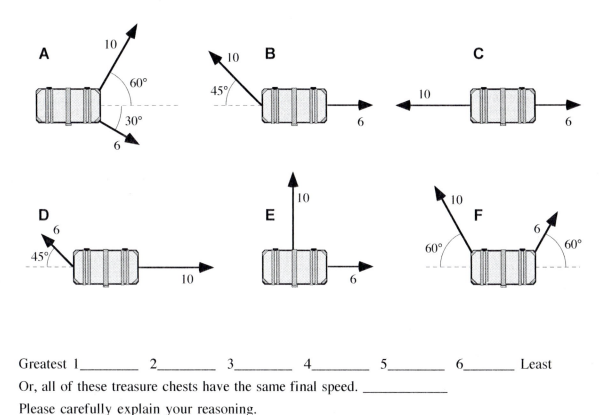

Greatest 1_____ 2_____ 3_____ 4_____ 5_____ 6_____ Least

Or, all of these treasure chests have the same final speed. _____

Please carefully explain your reasoning.

How sure were you of your ranking? (circle one)

Basically Guessed				Sure				Very Sure	
1	2	3	4	5	6	7	8	9	10

[17] K. W. Nicholson, D. Maloney, T. O'Kuma
Ranking Task Exercises in Physics: Student Edition   17

Two-Dimensional Forces on a Treasure Chest—Acceleration[18]

The six figures below represent treasure chests with two forces acting on them. We are looking down on the chests from above. The lengths of the force vectors represent the magnitude of the force. All chests have the same mass.

Rank these situations from greatest to least with regard to the resulting magnitude of acceleration of the treasure chest.

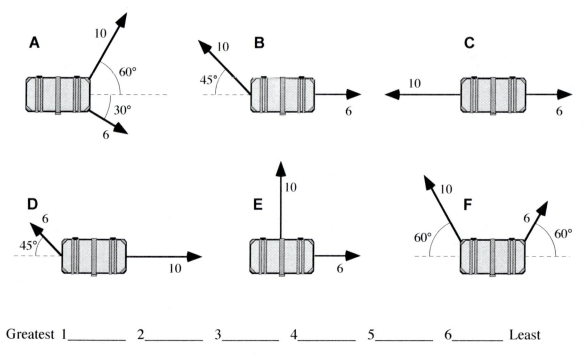

Greatest 1_____ 2_____ 3_____ 4_____ 5_____ 6_____ Least

Or, all of these treasure chests have the same magnitude of acceleration. _____

Please carefully explain your reasoning.

How sure were you of your ranking? (circle one)
Basically Guessed   Sure   Very Sure
1   2   3   4   5   6   7   8   9   10

[18] K.W. Nicholson, D. Maloney, T. O'Kuma
Ranking Task Exercises in Physics: Student Edition   18

Arrows—Acceleration[19]

The six figures below show arrows that have been shot into the air. All of the arrows were shot straight up and are the same size and shape. The arrows are made of different materials so they have different masses, and they have different speeds as they leave the bows. The values for each arrow are given in the figures. (We assume for this situation that the effect of air resistance can be neglected.) All start from the same height.

Rank these arrows, from greatest to least, on the basis of the acceleration of the arrows at the top of their flight.

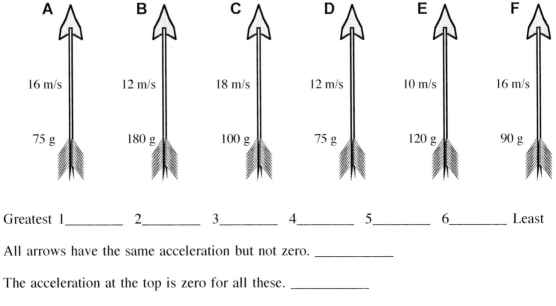

Greatest 1_____ 2_____ 3_____ 4_____ 5_____ 6_____ Least

All arrows have the same acceleration but not zero. _____

The acceleration at the top is zero for all these. _____

Please carefully explain your reasoning.

How sure were you of your ranking? (circle one)

Basically Guessed				Sure				Very Sure	
1	2	3	4	5	6	7	8	9	10

[19] T. O'Kuma, D. Maloney

Rocks Thrown Upward—Net Force[20]

Shown below are six rocks that have been thrown straight up into the air. The rocks all have the same shape, but they have different masses. The rocks are all thrown straight up, but at different speeds. The masses of the rocks and their speeds when released are given in the figures. (We assume for this situation that the effect of air resistance can be ignored.) All start from the same height.

Rank these rocks from greatest to least on the basis of the net force on the rocks after being thrown.

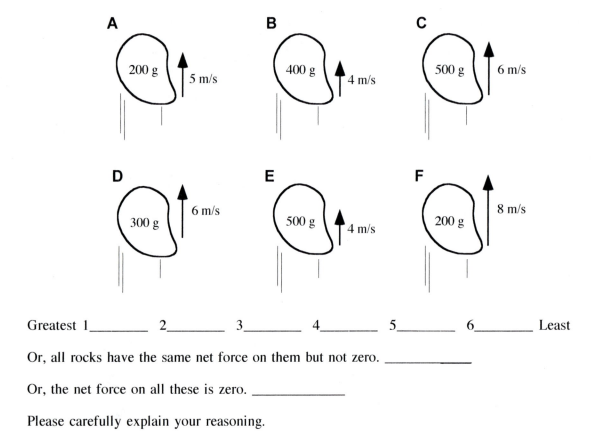

Greatest 1_____ 2_____ 3_____ 4_____ 5_____ 6_____ Least

Or, all rocks have the same net force on them but not zero. _____

Or, the net force on all these is zero. _____

Please carefully explain your reasoning.

How sure were you of your ranking? (circle one)
Basically Guessed   Sure   Very Sure
1   2   3   4   5   6   7   8   9   10

Rocks Thrown Downward—Net Force[21]

Shown below are six rocks that have been thrown straight down into the air. The rocks all have the same shape, but they have different masses. The rocks are all thrown straight down, but at different speeds. The masses of the rocks and their speeds when released are given in the figures. (We assume for this situation that the effect of air resistance can be ignored.) All start from the same height.

Rank these rocks from greatest to least on the basis of the magnitude of the net force on the rocks after being thrown down.

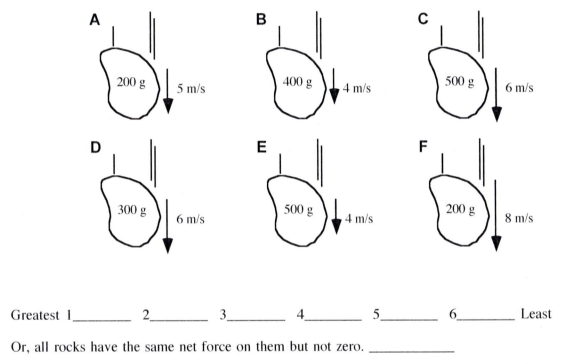

Greatest 1_____ 2_____ 3_____ 4_____ 5_____ 6_____ Least

Or, all rocks have the same net force on them but not zero. _____

Or, the net force on all these is zero. _____

Please carefully explain your reasoning.

How sure were you of your ranking? (circle one)

Basically Guessed   Sure   Very Sure

1   2   3   4   5   6   7   8   9   10

[21] T. O'Kuma, D. Maloney, C. Hieggelke

Ranking Task Exercises in Physics: Student Edition   21

Rocks Thrown Downward—Acceleration[22]

Shown below are six rocks that have been thrown straight down into the air. The rocks all have the same shape, but they have different masses. The rocks are all thrown straight down, but at different speeds. The masses of the rocks and their speeds when released are given in the figures. (We assume for this situation that the effect of air resistance can be ignored.) All start from the same height.

Rank these rocks from greatest to least on the basis of the magnitude of the acceleration of the rocks after being thrown.

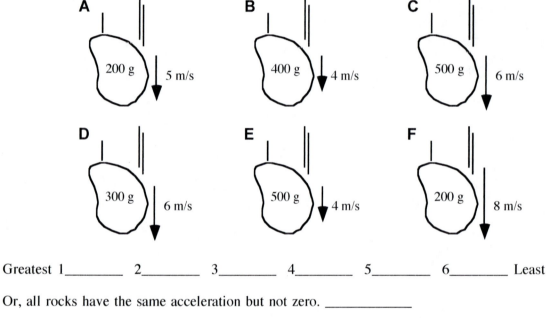

Greatest 1_____ 2_____ 3_____ 4_____ 5_____ 6_____ Least

Or, all rocks have the same acceleration but not zero. _____

Or, the acceleration of all these cases is zero. _____

Please carefully explain your reasoning.

How sure were you of your ranking? (circle one)
Basically Guessed   Sure   Very Sure
1   2   3   4   5   6   7   8   9   10

[22] T. O'Kuma, D. Maloney, C. Hieggelke

Model Rockets Moving Upward—Net Force[23]

The six figures below depict six model rockets that have just had their engines turned off. All of the rockets are aimed straight up, but their speeds differ. All of the rockets are the same size and shape, but they carry different loads, so their masses differ. The specific mass and speed for each rocket is given in each figure. (In this situation, we are going to ignore any effect air resistance may have on the rockets.) At the instant when the engines are turned off, the rockets are all at the same height.

Rank these model rockets, from greatest to least, on the basis of the net force on them after the engines have turned off.

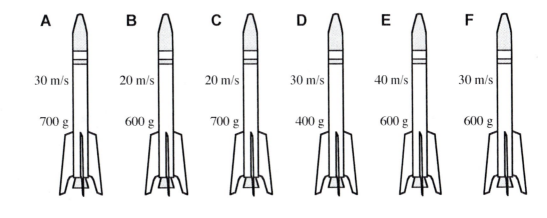

A	B	C	D	E	F
30 m/s	20 m/s	20 m/s	30 m/s	40 m/s	30 m/s
700 g	600 g	700 g	400 g	600 g	600 g

Greatest 1_____ 2_____ 3_____ 4_____ 5_____ 6_____ Least

Or, all rockets have the same net force on them (but not zero)._____

Or, the net force on all of these is zero. _____

Please carefully explain your reasoning.

How sure were you of your ranking? (circle one)
Basically Guessed   Sure   Very Sure
1   2   3   4   5   6   7   8   9   10

[23] T. O'Kuma

Blocks Attached to Fixed Objects—Rope Tension[24]

The eight figures below show various situations where blocks of different weights are attached by ropes to rigidly fixed objects or to other blocks, which are attached to fixed objects. The situations differ in a number of ways, as the figures show. The weights of the blocks are given in the figures, as well as the magnitudes and directions of any other forces that may be acting. Our interest is solely in the rope that is designated R in each figure.

Rank these arrangements, from greatest to least, on the basis of the tension in the rope R. That is, put first the arrangement where rope R is under the greatest tension and put last the arrangement where rope R is under the least tension.

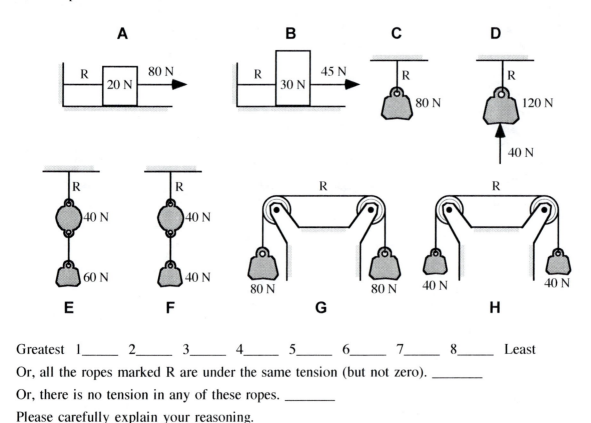

Greatest 1_____ 2_____ 3_____ 4_____ 5_____ 6_____ 7_____ 8_____ Least

Or, all the ropes marked R are under the same tension (but not zero). _____

Or, there is no tension in any of these ropes. _____

Please carefully explain your reasoning.

How sure were you of your ranking? (circle one)
Basically Guessed   Sure   Very Sure
1   2   3   4   5   6   7   8   9   10

Ball Motion Diagram—Net Force[25]

The following drawings indicate the motion of a ball subject to one or more forces on various surfaces from left to right. Each circle represents the position of the ball at succeeding instants of time. Each time-interval between successive positions is equal, and each ball has the same mass.

Rank the net force on the ball in each case from the highest to the lowest net force based on the figures. Assume the acceleration for each situation to be constant. Note: Positive is to the right and zero is greater than negative.

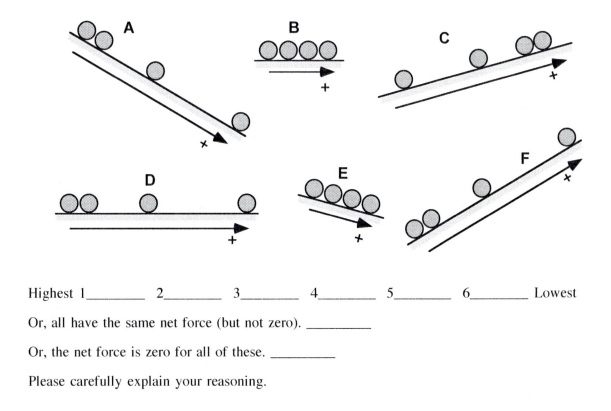

Highest 1_____ 2_____ 3_____ 4_____ 5_____ 6_____ Lowest

Or, all have the same net force (but not zero). _____

Or, the net force is zero for all of these. _____

Please carefully explain your reasoning.

How sure were you of your ranking? (circle one)

Basically Guessed				Sure				Very Sure	
1	2	3	4	5	6	7	8	9	10

[25] D. Maloney, T. O'Kuma

Force Acceleration Graphs—Mass[26]

The following graphs plot force vs. acceleration for several objects. All graphs have the same scale for each respective axis.

Rank each situation according to mass. That is, order the situations from the largest to the smallest mass that the force is acting upon.

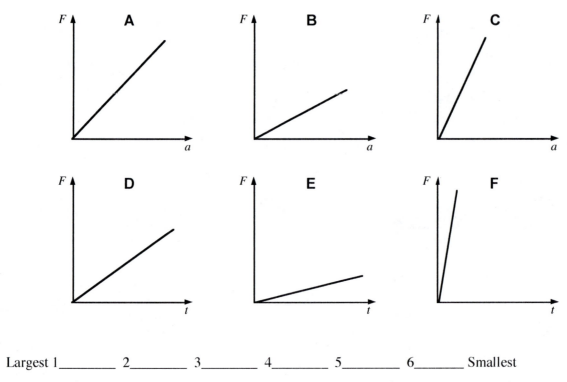

Largest 1_____ 2_____ 3_____ 4_____ 5_____ 6_____ Smallest

Or, all the masses are the same. _____
Please carefully explain your reasoning.

How sure were you of your ranking? (circle one)
Basically Guessed   Sure   Very Sure
1   2   3   4   5   6   7   8   9   10

---

[26] D. Schramme, C. Fang, B. Speers

Two Different Blocks and a Pulley—Tension[27]

Each figure below shows two blocks hanging from the ends of a strong but massless string that passes over a massless, frictionless pulley. In each figure, the block on the left is more massive than the block on the right, so the block on the left accelerates down, and the block on the right accelerates up. The mass of each block is given in the figures.

Rank the figures from greatest to least on the basis of the tension in the string for the system of blocks.

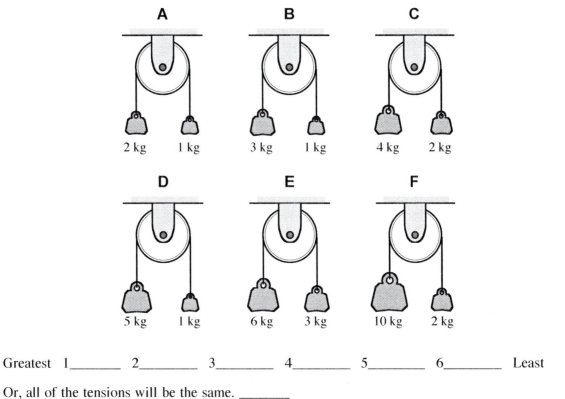

A	B	C
2 kg 1 kg	3 kg 1 kg	4 kg 2 kg

D	E	F
5 kg 1 kg	6 kg 3 kg	10 kg 2 kg

Greatest 1_____ 2_____ 3_____ 4_____ 5_____ 6_____ Least

Or, all of the tensions will be the same. _____

Please carefully explain your reasoning.

How sure were you of your ranking? (circle one)

Basically Guessed   Sure   Very Sure

1   2   3   4   5   6   7   8   9   10

[27] S. Loucks, D. Maloney, T. O'Kuma

Ranking Task Exercises in Physics: Student Edition   27

Ropes Pulling Boxes—Acceleration[28]

The figures below show boxes that are being pulled by ropes along frictionless surfaces, accelerating toward the left. All of the boxes are identical. The pulling force applied to the left-most rope is the same in each figure. As you can see, some of the boxes are pulled by ropes attached to the box in front of them.

Rank the masses from greatest to least on the basis of the acceleration of the masses.

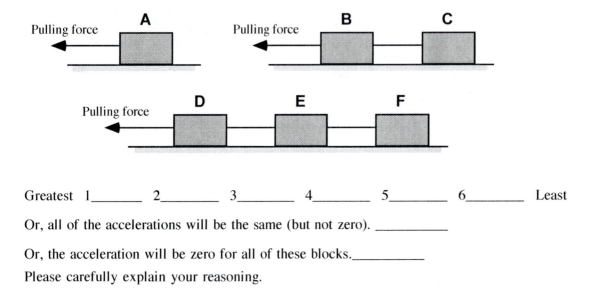

Greatest 1_____ 2_____ 3_____ 4_____ 5_____ 6_____ Least

Or, all of the accelerations will be the same (but not zero). _____

Or, the acceleration will be zero for all of these blocks._____

Please carefully explain your reasoning.

How sure were you of your ranking? (circle one)
Basically Guessed   Sure   Very Sure
1   2   3   4   5   6   7   8   9   10

[28] S. Loucks
Ranking Task Exercises in Physics: Student Edition   28

Ropes Pulling Boxes—Rope Tension[29]

The figures below show boxes that are being pulled by ropes along frictionless surfaces, accelerating toward the left. All of the boxes are identical, and the acceleration is the same in each figure. As you can see, some of the boxes are pulled by ropes attached to the box in front of them.

Rank the ropes from greatest to least on the basis of the tension in the rope.

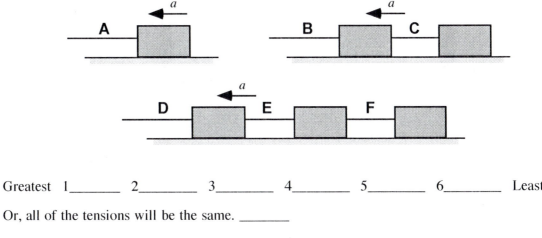

Greatest 1_____ 2_____ 3_____ 4_____ 5_____ 6_____ Least

Or, all of the tensions will be the same. _____

Please carefully explain your reasoning.

How sure were you of your ranking? (circle one)
Basically Guessed   Sure   Very Sure
1   2   3   4   5   6   7   8   9   10

Two Different Blocks and a Pulley—Net Force[30]

Each figure below shows two blocks hanging from the ends of a strong but massless string, which passes over a massless, frictionless pulley. In each figure, the block on the left is more massive than the block on the right, so the block on the left accelerates down, and the block on the right accelerates up. The mass of each block is given in the figures.

Rank the figures from greatest to least on the basis of the net force that accelerates the system of blocks.

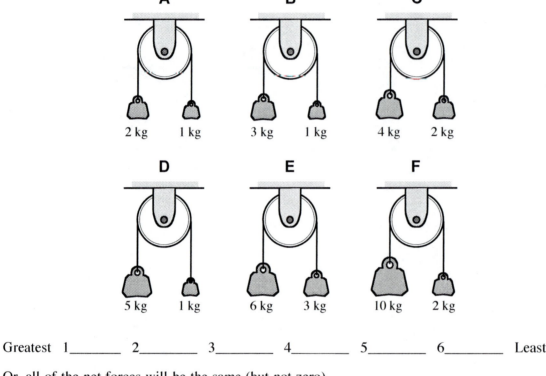

A	B	C
2 kg 1 kg	3 kg 1 kg	4 kg 2 kg

D	E	F
5 kg 1 kg	6 kg 3 kg	10 kg 2 kg

Greatest 1_____ 2_____ 3_____ 4_____ 5_____ 6_____ Least

Or, all of the net forces will be the same (but not zero). _____

Or, the net force is zero for all of these. _____

Please carefully explain your reasoning.

How sure were you of your ranking? (circle one)
Basically Guessed   Sure   Very Sure
1   2   3   4   5   6   7   8   9   10

Moving Car and Boat Trailer—Force Difference[31]

In the six figures below, all the boat trailers and cars are identical but the boat trailers have different loads.

Rank from greatest to least on the basis of the difference between the strength (magnitude) of the force the car exerts on the boat trailer, and the strength of the force the boat trailer exerts on the car.

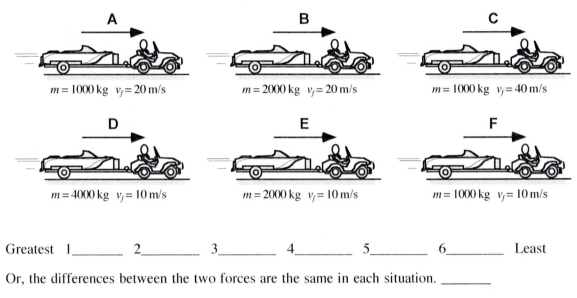

A
$m = 1000\,kg \quad v_f = 20\,m/s$

B
$m = 2000\,kg \quad v_f = 20\,m/s$

C
$m = 1000\,kg \quad v_f = 40\,m/s$

D
$m = 4000\,kg \quad v_f = 10\,m/s$

E
$m = 2000\,kg \quad v_f = 10\,m/s$

F
$m = 1000\,kg \quad v_f = 10\,m/s$

Greatest 1_____ 2_____ 3_____ 4_____ 5_____ 6_____ Least

Or, the differences between the two forces are the same in each situation. _____

Please carefully explain your reasoning.

How sure were you of your ranking? (circle one)

Basically Guessed   Sure   Very Sure
1   2   3   4   5   6   7   8   9   10

[31] P. Golden, A. Dickison, D. Maloney, T. O'Kuma, C. Hieggelke
Ranking Task Exercises in Physics: Student Edition   31

Accelerating Car and Boat Trailer—Force Difference[32]

In the six figures below, all the trailers and cars are identical but the boat trailers have different loads.

Rank from greatest to least on the basis of the difference between the strength (magnitude) of the force the car exerts on the boat trailer and the strength of the force the trailer exerts on the car during the period when the boat trailers are accelerating from rest to the given final speeds.

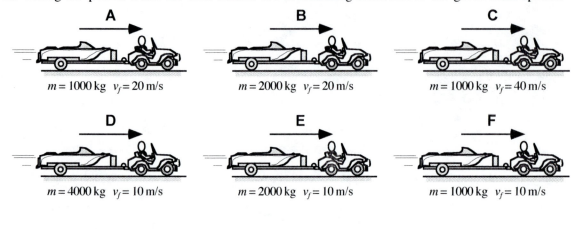

A
$m = 1000\,\text{kg}$ $v_f = 20\,\text{m/s}$

B
$m = 2000\,\text{kg}$ $v_f = 20\,\text{m/s}$

C
$m = 1000\,\text{kg}$ $v_f = 40\,\text{m/s}$

D
$m = 4000\,\text{kg}$ $v_f = 10\,\text{m/s}$

E
$m = 2000\,\text{kg}$ $v_f = 10\,\text{m/s}$

F
$m = 1000\,\text{kg}$ $v_f = 10\,\text{m/s}$

Greatest 1_____ 2_____ 3_____ 4_____ 5_____ 6_____ Least

Or, the differences between the two forces are the same in all situations. _____

Please carefully explain your reasoning.

How sure were you of your ranking? (circle one)
Basically Guessed   Sure   Very Sure
1   2   3   4   5   6   7   8   9   10

[32] P. Golden, A. Dickison, D. Maloney, T. O'Kuma, C. Hieggelke

Car and Boat Trailer on an Incline—Force Difference[33]

In the following six figures, all the cars are identical, but the boat trailers have different loads. The velocities and accelerations vary but all inclines are the same.

Rank from greatest to least on the basis of the difference between the strength (magnitude) of the force the car exerts on the boat trailer and the strength of the force the boat trailer exerts on the car.

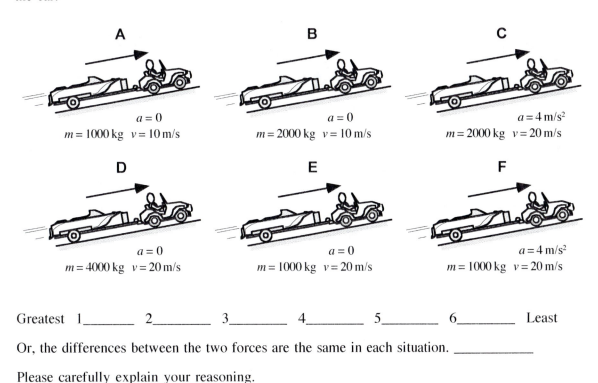

A
$a = 0$
$m = 1000\,\text{kg}$ $v = 10\,\text{m/s}$

B
$a = 0$
$m = 2000\,\text{kg}$ $v = 10\,\text{m/s}$

C
$a = 4\,\text{m/s}^2$
$m = 2000\,\text{kg}$ $v = 20\,\text{m/s}$

D
$a = 0$
$m = 4000\,\text{kg}$ $v = 20\,\text{m/s}$

E
$a = 0$
$m = 1000\,\text{kg}$ $v = 20\,\text{m/s}$

F
$a = 4\,\text{m/s}^2$
$m = 1000\,\text{kg}$ $v = 20\,\text{m/s}$

Greatest 1_____ 2_____ 3_____ 4_____ 5_____ 6_____ Least

Or, the differences between the two forces are the same in each situation. _____

Please carefully explain your reasoning.

How sure were you of your ranking? (circle one)
Basically Guessed   Sure   Very Sure
1   2   3   4   5   6   7   8   9   10

[33] P. Golden, A. Dickison, D. Maloney, T. O'Kuma, C. Hieggelke
Ranking Task Exercises in Physics: Student Edition   33

Forces on Objects on Smooth Surfaces—Velocity Change[34]

All objects start at rest. Two forces act on an object that is on a frictionless surface, as shown below.

Rank these situations from greatest change in velocity to least change in velocity. (Note: All vectors directed to the right are positive, and those to the left are negative. Also, 0 m/s > -10 m/s).

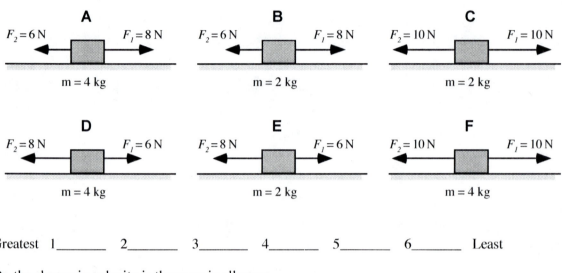

Greatest 1_____ 2_____ 3_____ 4_____ 5_____ 6_____ Least

Or, the change in velocity is the same in all cases. _____

Or, the velocity will not change in any of these situations. _____

Please carefully explain your reasoning.

How sure were you of your ranking? (circle one)
Basically Guessed   Sure   Very Sure
1   2   3   4   5   6   7   8   9   10

_____

[34] R. Krupp

Forces on Objects on Smooth Surfaces—Speed Change[35]

All objects start at rest. Two forces act on an object that is on a frictionless surface, as shown below.

Rank these situations from greatest change in speed to least change in speed.

A
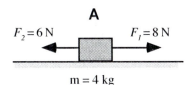

B
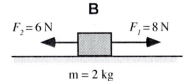

C
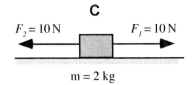

D
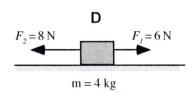

E
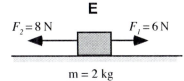

F
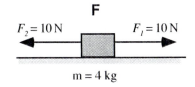

Greatest 1_____ 2_____ 3_____ 4_____ 5_____ 6_____ Least

Or, the change in speed is the same in all cases. _____

Or, the speed does not change for any of these cases. _____

Please carefully explain your reasoning.

How sure were you of your ranking? (circle one)
Basically Guessed   Sure   Very Sure
1   2   3   4   5   6   7   8   9   10

[35] R. Krupp

Forces on Objects on Rough Surfaces—Velocity Change[36]

Two forces act on identical objects that are on rough surfaces, as shown below. The forces of maximum static and kinetic friction for each case are both 1 N. All objects start at rest.

Rank these situations from greatest change in velocity to least change in velocity. (Note: All vectors directed to the right are positive, and those to the left are negative. Also, 0 m/s > -10 m/s.)

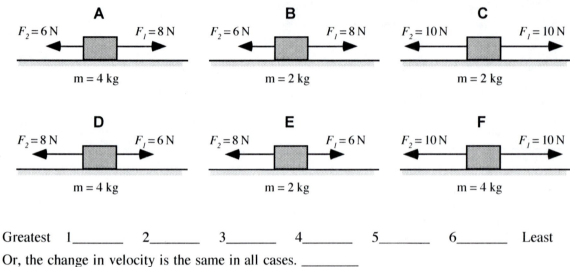

Greatest   1_____   2_____   3_____   4_____   5_____   6_____   Least

Or, the change in velocity is the same in all cases. _____

Please carefully explain your reasoning.

How sure were you of your ranking? (circle one)

Basically Guessed   Sure   Very Sure

1   2   3   4   5   6   7   8   9   10

[36] R. Krupp

Forces on Objects on Rough Surfaces—Speed Change[37]

Two forces act on identical objects that are on rough surfaces, as shown below. The forces of maximum static and kinetic friction for each case are both 2 N. All objects start at rest.

Rank these situations from greatest change in speed to least change in speed.

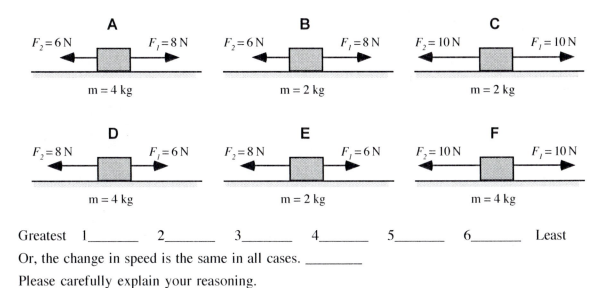

Greatest   1_____   2_____   3_____   4_____   5_____   6_____   Least

Or, the change in speed is the same in all cases. _____

Please carefully explain your reasoning.

How sure were you of your ranking? (circle one)

Basically Guessed   Sure   Very Sure

1   2   3   4   5   6   7   8   9   10

[37] R. Krupp

Ranking Task Exercises in Physics: Student Edition   37

Person in an Elevator Moving Upward—Scale Weight[38]

The figures below depict situations where a person is standing on a scale in six identical elevators. Each person weighs 600 N when the elevators are stationary. Each elevator now moves (accelerates) according to the specified arrow that is drawn next to it. In all cases where the elevator is moving, it is moving <u>upward</u>.

Rank the figures, from greatest to least, on the basis of the *scale weight* of each person as registered on each scale. (Use $g = 9.8$ m/s^2).

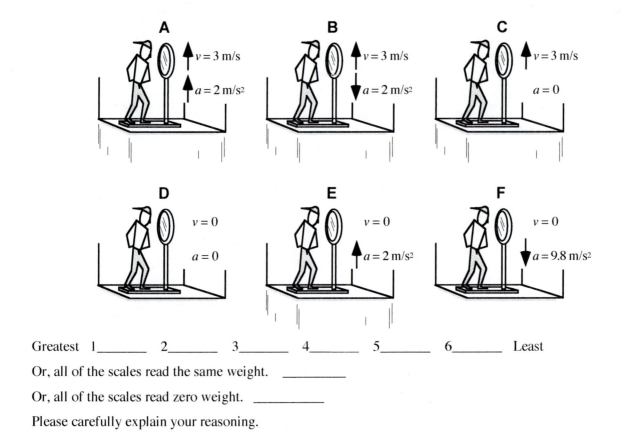

A $v = 3$ m/s $a = 2$ m/s^2

B $v = 3$ m/s $a = 2$ m/s^2

C $v = 3$ m/s $a = 0$

D $v = 0$ $a = 0$

E $v = 0$ $a = 2$ m/s^2

F $v = 0$ $a = 9.8$ m/s^2

Greatest 1_____ 2_____ 3_____ 4_____ 5_____ 6_____ Least

Or, all of the scales read the same weight. _____

Or, all of the scales read zero weight. _____

Please carefully explain your reasoning.

How sure were you of your ranking? (circle one)
Basically Guessed   Sure   Very Sure
1   2   3   4   5   6   7   8   9   10

---

[38] O. Karmon

Person in an Elevator Moving Downward—Scale Weight[39]

The figures below depict situations where a person is standing on a scale in six identical elevators. Each person weighs 600 N when the elevators are stationary. Each elevator now moves (accelerates) according to the specified arrow that is drawn next to it. In all cases where the elevator is moving, it is moving <u>downward</u>.

Rank the figures, from greatest to least, on the basis of the *scale weight* of each person as registered on each scale. (Use $g = 9.8$ m/s².)

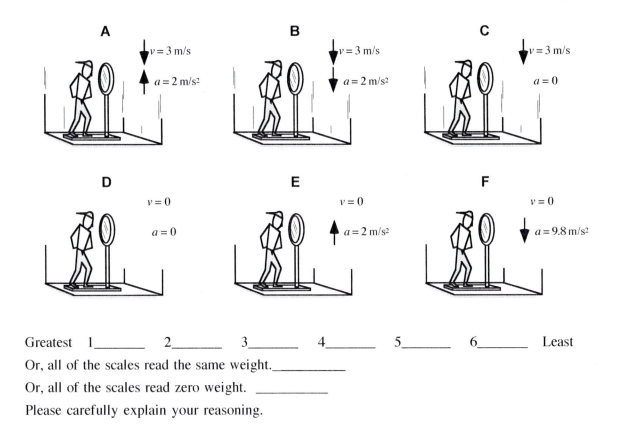

Greatest   1_____   2_____   3_____   4_____   5_____   6_____   Least

Or, all of the scales read the same weight._____

Or, all of the scales read zero weight. _____

Please carefully explain your reasoning.

How sure were you of your ranking? (circle one)

Basically Guessed				Sure				Very Sure	
1	2	3	4	5	6	7	8	9	10

Two Blocks at Rest—Force Difference[40]

Shown below are six arrangements of two wooden blocks. There are two different mass blocks, either 100 g or 200 g. In all of the arrangements, the blocks are in contact, that is, they are touching each other. Also, in all of the arrangements the blocks are at rest, i.e., they are not moving. As you can see, one of the blocks given in each arrangement is labeled **A**, and the other is labeled **B**. The mass of each block is given in the figures.

Rank these arrangements from largest to smallest on the basis of the difference of the strengths (magnitudes) of the forces between the force **A** exerts on **B** and the force **B** exerts on **A**. In other words, the arrangement where the force **A** exerts on **B** minus the force **B** exerts on **A** is the largest will rank first. In the same way, the arrangement where the force **A** exerts on **B** minus the force **B** exerts on **A** is the smallest will rank last. Keep in mind that some of these values might be negative. If **B** is exerting a stronger force on **A** than **A** exerts on **B**, then the difference will be negative. Negative values are smaller than positive values or zero. (A force is a push or a pull.)

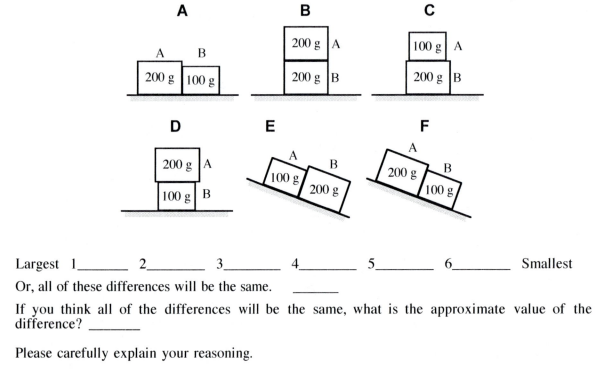

Largest 1_____ 2_____ 3_____ 4_____ 5_____ 6_____ Smallest

Or, all of these differences will be the same. _____

If you think all of the differences will be the same, what is the approximate value of the difference? _____

Please carefully explain your reasoning.

How sure were you of your ranking? (circle one)
Basically Guessed   Sure   Very Sure
1   2   3   4   5   6   7   8   9   10

Two Moving Blocks—Force Difference[41]

Shown below are six arrangements of two wooden blocks both moving left to right at 2 m/s. There are two different mass blocks, either 100 g or 200 g. In all of the arrangements, the blocks are in contact, that is, they are touching each other. As you can see, one of the blocks given in each arrangement is labeled **A**, and the other is labeled **B**. The mass of each block is given in the figures.

Rank these arrangements from largest to smallest on the basis of the difference of the strengths (magnitudes) of the forces between the force **A** exerts on **B** and the force **B** exerts on **A**. In other words, the arrangement where the force **A** exerts on **B** minus the force **B** exerts on **A** is the largest will rank first. In the same way the arrangement where the force **A** exerts on **B** minus the force **B** exerts on **A** is the smallest will rank last. Keep in mind that some of these values might be negative. If **B** is exerting a stronger force on **A** than **A** exerts on **B**, then the difference will be negative. Negative values are smaller than positive values or zero. (A force is a push or a pull.)

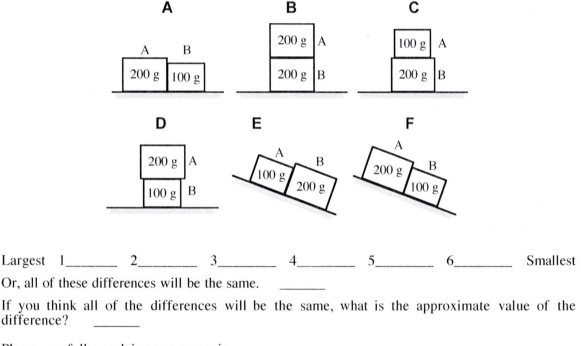

Largest 1_____ 2_____ 3_____ 4_____ 5_____ 6_____ Smallest

Or, all of these differences will be the same. _____

If you think all of the differences will be the same, what is the approximate value of the difference? _____

Please carefully explain your reasoning.

How sure were you of your ranking? (circle one)
Basically Guessed   Sure   Very Sure
1   2   3   4   5   6   7   8   9   10

[41] D. Maloney, C. Hieggelke
Ranking Task Exercises in Physics: Student Edition   41

Two Accelerating Blocks—Force Difference[42]

Shown below are six arrangements of two wooden blocks both moving left to right at 2 m/s and accelerating in the same direction at 3 m/s^2. There are two different mass blocks, either 100 g or 200 g. In all of the arrangements, the blocks are in contact, that is, they are touching each other. As you can see, one of the blocks given in each arrangement is labeled **A**, and the other is labeled **B**. The mass of each block is given in the figures.

Rank these arrangements from largest to smallest on the basis of the difference of the strengths (magnitudes) of the forces between the force **A** exerts on **B** and the force **B** exerts on **A**. In other words, the arrangement where the force **A** exerts on **B** minus the force **B** exerts on **A** is the largest will rank first. In the same way the arrangement where the force **A** exerts on **B** minus the force **B** exerts on **A** is the smallest will rank last. Keep in mind that some of these values might be negative. If **B** is exerting a stronger force on **A** than **A** exerts on **B**, then the difference will be negative. Negative values are smaller than positive values or zero. (A force is a push or a pull.)

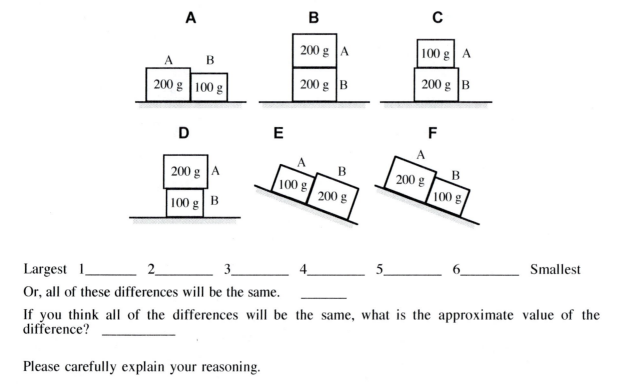

Largest 1_____ 2_____ 3_____ 4_____ 5_____ 6_____ Smallest

Or, all of these differences will be the same.   _____

If you think all of the differences will be the same, what is the approximate value of the difference? _____

Please carefully explain your reasoning.

How sure were you of your ranking? (circle one)

Basically Guessed   Sure   Very Sure
1   2   3   4   5   6   7   8   9   10

[42] D. Maloney, C. Hieggelke

Ranking Task Exercises in Physics: Student Edition   42

Horizontal Arrows at Different Distances—Force[43]

The figures below show arrows that have been shot from bows. All of the arrows are identical, and they are moving horizontally to the right. The arrows are at different points in their paths from the bows to the targets. The distances the arrows have traveled in reaching the points shown are given in the figures. Also given in the figures are the speeds the arrows have at the points shown.

Rank these situations, from greatest to least, on the basis of the rightward pointing force, i.e., the force acting in the direction the arrow is moving, acting on each arrow at the point shown. That is, put first the arrow with the largest horizontal force acting on it, and put last the arrow with the smallest horizontal force. (A force is a push or pull.) We are ignoring any effects of air in these situations.

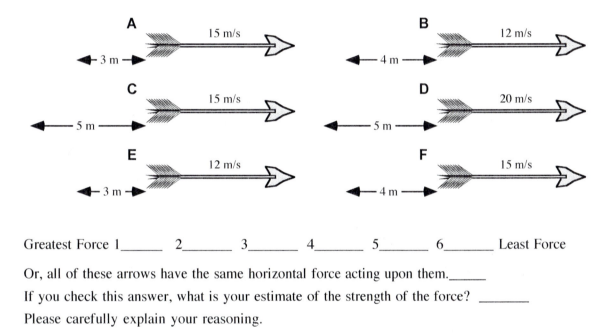

Greatest Force 1_____ 2_____ 3_____ 4_____ 5_____ 6_____ Least Force

Or, all of these arrows have the same horizontal force acting upon them._____

If you check this answer, what is your estimate of the strength of the force? _____

Please carefully explain your reasoning.

How sure were you of your ranking? (circle one)
Basically Guessed				Sure				Very Sure	
1	2	3	4	5	6	7	8	9	10

[43] D. Maloney
Ranking Task Exercises in Physics: Student Edition   43

Horizontal Arrows at Different Times—Force[44]

The figures below show arrows that have been shot from bows. All of the arrows are identical, and they are moving horizontally to the right. The arrows are at different points in their paths from the bows to the targets. The times since being shot vary for the arrows. These times are given in the figures. Also given in the figures are the speeds the arrows have at the specified times.

Rank these situations, from greatest to least, on the basis of the rightward pointing force; i.e., the force acting in the direction the arrow is moving, acting on each arrow at the point shown. That is, put first the arrow with the largest horizontal force acting on it, and put last the arrow with the smallest horizontal force. (A force is a push or pull.) We are ignoring any effects of air in these situations.

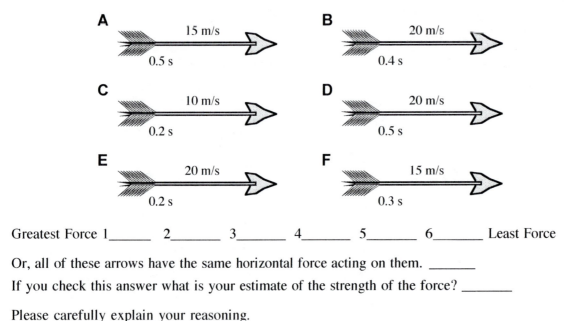

Greatest Force 1_____ 2_____ 3_____ 4_____ 5_____ 6_____ Least Force

Or, all of these arrows have the same horizontal force acting on them. _____

If you check this answer what is your estimate of the strength of the force? _____

Please carefully explain your reasoning.

How sure were you of your ranking? (circle one)

Basically Guessed				Sure				Very Sure	
1	2	3	4	5	6	7	8	9	10

[44] D. Maloney

Horizontal Arrows at Different Distances and Times—Force[45]

The figures below show arrows that have been shot from bows. All of the arrows are identical, and they are moving horizontally to the right. The arrows are at different points in their paths from the bows to the targets. The distances the arrows have traveled and the times required are given in the figures.

Rank these situations, from greatest to least, on the basis of the rightward pointing force; i.e., the force acting in the direction the arrow is moving, acting on each arrow at the point shown. That is, put first the arrow with the largest horizontal force acting on it, and put last the arrow with the smallest horizontal force. (A force is a push or pull.) We are ignoring any effects of air in these situations.

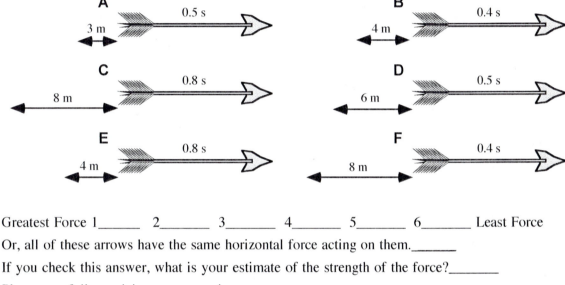

Greatest Force 1_____ 2_____ 3_____ 4_____ 5_____ 6_____ Least Force

Or, all of these arrows have the same horizontal force acting on them._____

If you check this answer, what is your estimate of the strength of the force?_____

Please carefully explain your reasoning.

How sure were you of your ranking? (circle one)
Basically Guessed   Sure   Very Sure
1   2   3   4   5   6   7   8   9   10

[45] D. Maloney

Water Over a Waterfall—Time to Reach Ground[46]

Pictured below are six waterfalls, all of which have the same amount of water flowing over them. The waterfalls differ in height and in the speed of the water as it goes over the edge. The specific values of the heights and speeds are given in the figures.

Rank these situations from longest to shortest based on how long it takes the water to go from the top of the falls to the bottom. That is, put first the situation where it takes the water the most time to go from the top of the falls to the bottom, and put last the one that takes the least time.

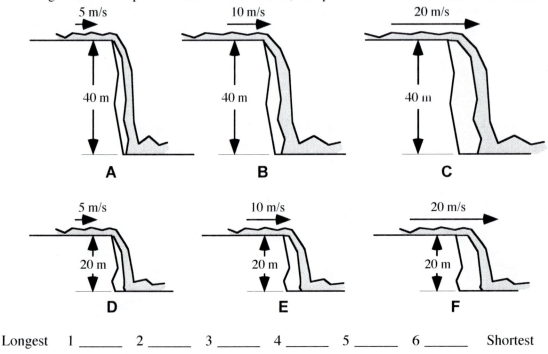

Longest   1 _____   2 _____   3 _____   4 _____   5 _____   6 _____   Shortest

Or, water from all of the waterfalls reaches the bottom at the same time. _____

Please carefully explain your reasoning.

How sure were you of your ranking? (circle one)

Basically Guessed				Sure				Very Sure	
1	2	3	4	5	6	7	8	9	10

[46] D. Maloney

Rifle Shots—Time to Hit Ground[47]

The six figures below show rifles that are being fired horizontally, i.e., straight out, off platforms. The bullets fired from the rifles are all identical, but the rifles propel the bullets at different speeds. The specific speed of each bullet and the height of each platform are given. All of the bullets miss the targets and hit the ground.

Rank these bullets, from longest to shortest, on the basis of how long it takes a bullet to hit the ground. That is, put first the bullet that will take the longest time from being fired to hitting the ground, and put last the bullet that will take the shortest time.

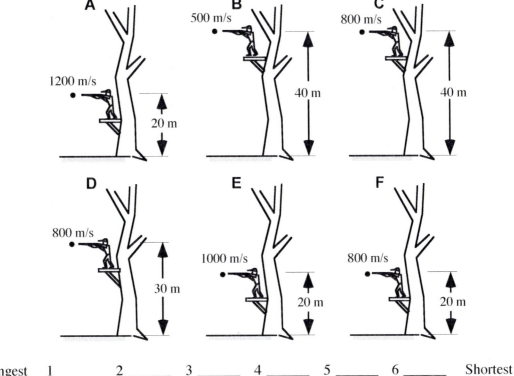

Longest   1 _____   2 _____   3 _____   4 _____   5 _____   6 _____   Shortest

Or, all of the bullets reach the ground at the same. _____

Please carefully explain your reasoning.

How sure were you of your ranking? (circle one)

Basically Guessed				Sure				Very Sure	
1	2	3	4	5	6	7	8	9	10

Toy Trucks Rolling Off Tables—Time in Air[48]

Shown in the six figures below are toy trucks that are rolling along on tables. All of the trucks are identical. The tables they are on vary in height, and the speeds of the trucks vary. Specific values for the heights and speeds are given in the figures. Each truck rolls off the end of its table.

Rank these trucks from longest to shortest, based on how long the trucks are in the air. That is, put first the truck that takes the longest time to go from the edge of the table to the ground, and put last the truck that takes the least time.

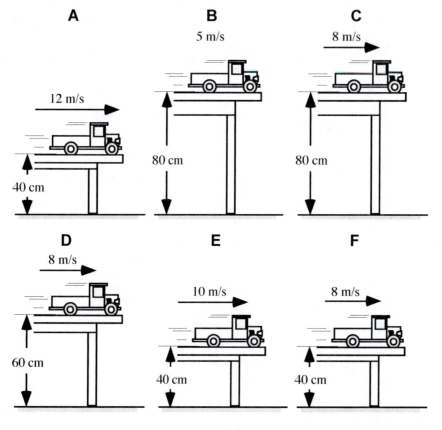

Longest   1 _____   2 _____   3 _____   4 _____   5 _____   6 _____   Shortest

Or, all of the toy trucks are in the air the same amount of time. _____

Please carefully explain your reasoning.

How sure were you of your ranking? (circle one)
Basically Guessed   Sure   Very Sure
1   2   3   4   5   6   7   8   9   10

Arrows—Maximum Heights[49]

The six figures below show arrows that have been shot into the air. All of the arrows were shot at the same angle and are the same size and shape. The arrows are made of different materials so they have different masses, and they have different speeds as they leave the bows. The values for each arrow are given in the figures. (We assume for this situation that the effect of air resistance can be ignored.) All start from the same height.

Rank these arrows from greatest to least on the basis of the maximum heights the arrows reach.

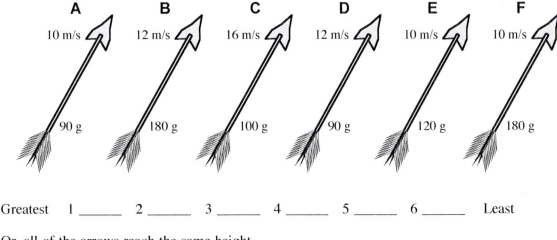

Greatest   1 _____   2 _____   3 _____   4 _____   5 _____   6 _____   Least

Or, all of the arrows reach the same height. _____

Please carefully explain your reasoning.

How sure were you of your ranking? (circle one)

Basically Guessed				Sure				Very Sure	
1	2	3	4	5	6	7	8	9	10

[49] D. Maloney
Ranking Task Exercises in Physics: Student Edition   49

Rock Throw—Maximum Heights[50]

Shown below are six rocks that have been thrown into the air. The rocks all have the same size and shape, but they have different masses. The rocks are all thrown at the same angle, but at different speeds. The masses of the rocks and their speeds, when released, are given in the figures. (We assume for this situation that the effect of air resistance can be ignored.) All start from the same height.

Rank these rocks from greatest to least on the basis of the maximum heights the rocks reach.

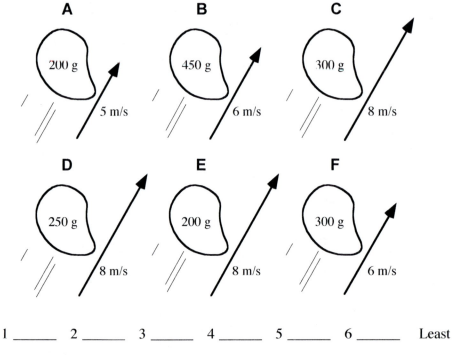

Greatest   1 _____ 2 _____ 3 _____ 4 _____ 5 _____ 6 _____   Least

Or, all rocks reach the same height. _____

Please carefully explain your reasoning.

How sure were you of your ranking? (circle one)

Basically Guessed   Sure   Very Sure

1   2   3   4   5   6   7   8   9   10

[50] D. Maloney

Model Rockets Fired at an Angle—Horizontal Speed at Top[51]

The six figures below depict six model rockets that have just had their engines turned off. All of the rockets are aimed upward at the same angle, but their speeds differ. All of the rockets are the same size and shape, but they carry different loads, so their masses differ. The specific mass and speed for each rocket is given in each figure. (In this situation we are going to ignore any effect air resistance may have on the rockets.) At the instant when the engines are turned off, the rockets are all at the same height.

Rank these model rockets from greatest to least on the basis of the horizontal speed at the top (at the maximum height).

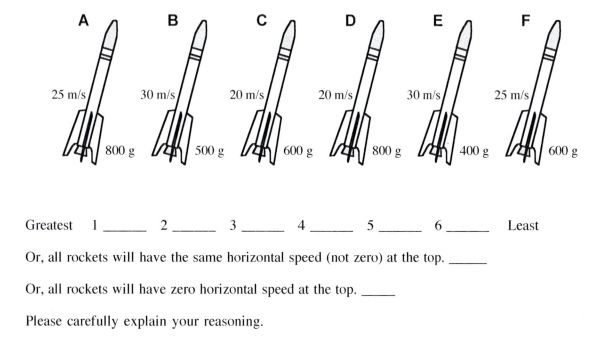

Greatest   1 _____   2 _____   3 _____   4 _____   5 _____   6 _____   Least

Or, all rockets will have the same horizontal speed (not zero) at the top. _____

Or, all rockets will have zero horizontal speed at the top. _____

Please carefully explain your reasoning.

How sure were you of your ranking? (circle one)
Basically Guessed   Sure   Very Sure
1   2   3   4   5   6   7   8   9   10

Cannon Shots—Acceleration at the Top[52]

The figures below depict six cannons shooting shells into the air. All of the cannons are aimed at the same angle of 35 degrees. All of the cannons are identical. The shells are all the same size and shape, but the masses of the shells, as well as their speeds as they leave the cannons, are different. The values of these variables are specified in the figures.

Rank these situations from highest to lowest on the basis of the acceleration at the highest point reached by the shells. (We assume for this situation that the effect of air resistance can be ignored.)

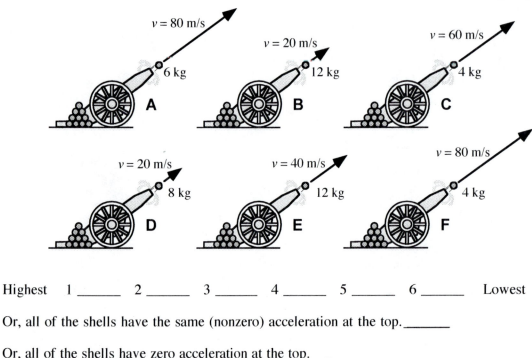

Highest   1 _____   2 _____   3 _____   4 _____   5 _____   6 _____   Lowest

Or, all of the shells have the same (nonzero) acceleration at the top. _____

Or, all of the shells have zero acceleration at the top. ___

Please carefully explain your reasoning.

How sure were you of your ranking? (circle one)

Basically Guessed				Sure				Very Sure	
1	2	3	4	5	6	7	8	9	10

Carts on Incline—Height After Leaving Incline[53]

Shown below are diagrams of situations where carts, initially moving along a horizontal surface, run up onto short inclines. The carts go up off the inclines. All of the inclines have the same angle, and they are very smooth, so there is essentially no friction. The masses of the carts and their speeds when they reach the inclines are given in the figures.

Rank the situations in order of how high the carts go after leaving the incline. That is, order the situations from the one where the cart goes highest (greatest vertical distance) to that where the cart goes lowest.

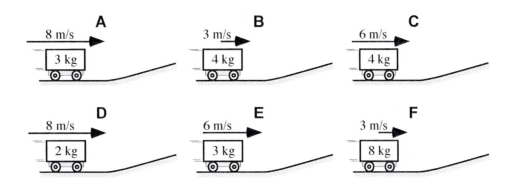

Highest   1 _____   2 _____   3 _____   4 _____   5 _____   6 _____   Lowest

Or, all of the carts reach the same height. _____

Please carefully explain your reasoning.

How sure were you of your ranking? (circle one)

Basically Guessed				Sure				Very Sure	
1	2	3	4	5	6	7	8	9	10

[53] D. Maloney

Projectile—Horizontal Distance[54]

The pictures below depict cannonballs of two different masses projected upward and forward. The cannonballs are projected at various angles above the horizontal, but all are projected with the same horizontal component of velocity.

Rank these situations from largest to smallest according to the horizontal distance traveled by the balls.

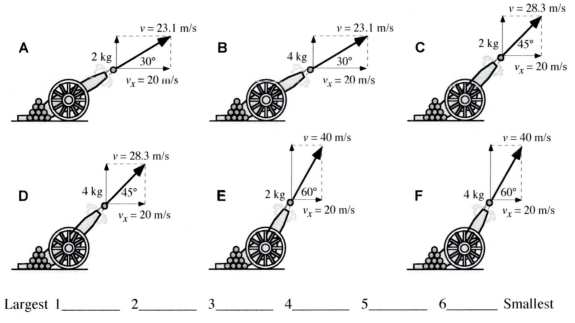

Largest 1_____ 2_____ 3_____ 4_____ 5_____ 6_____ Smallest

All distances traveled are the same. _____

Please carefully explain your reasoning.

How sure were you of your ranking? (circle one)

Basically Guessed				Sure				Very Sure	
1	2	3	4	5	6	7	8	9	10

[54] S. Heath

Projectile—Time in Air[55]

The pictures below depict cannonballs of two different masses projected upward and forward. The cannonballs are projected at various angles above the horizontal, but all are projected with the same vertical component of velocity.

Rank these situations from largest to smallest according to the time the balls are in the air.

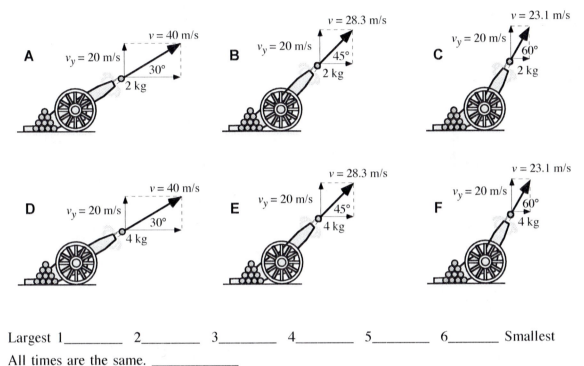

Largest 1_____ 2_____ 3_____ 4_____ 5_____ 6_____ Smallest

All times are the same. _____

Please carefully explain your reasoning.

How sure were you of your ranking? (circle one)
Basically Guessed   Sure   Very Sure
1   2   3   4   5   6   7   8   9   10

Boxcars and Ropes—Stopping Force in Same Distance[56]

In a western movie, bandits stopped a runaway boxcar carrying gold by using many ropes tied to trees. Given below are six boxcars that are moving along horizontal railroads at specified speeds. Also given are the masses of the boxcars. All of the boxcars are the same size and shape, but they are carrying loads with different masses. All of these boxcars are going to be stopped by plowing through a large number of these secured ropes. All of the boxcars need to be stopped in the same distance.

Rank these situations from greatest to least on the basis of the strength of the forces that will be needed to stop the boxcars in the same distance. That is, put first the boxcar on which the strongest force will have to be applied to stop it in x meters, and put last the boxcar on which the weakest force will be applied to stop the boxcar in the same distance.

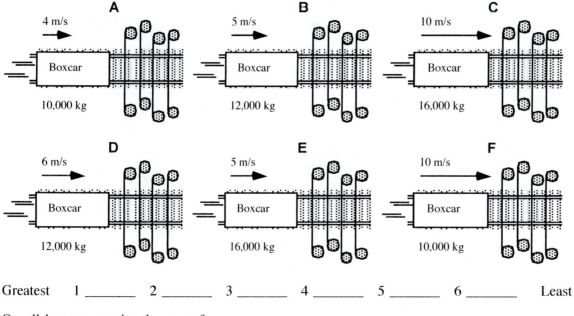

Greatest   1 _____ 2 _____ 3 _____ 4 _____ 5 _____ 6 _____   Least

Or, all boxcars require the same force. _____

Please carefully explain your reasoning.

How sure were you of your ranking? (circle one)

Basically Guessed				Sure				Very Sure	
1	2	3	4	5	6	7	8	9	10

Cars and Barriers—Stopping Force in Same Distance I[57]

Given below are six cars that are moving along horizontal roads at specified speeds. Also given are the masses of the cars. All of the cars are the same size and shape, but they are carrying loads with different masses. All of these cars are going to be stopped by plowing into barrel barriers. All of the cars are going to be stopped in the same distance.

Rank these situations from greatest to least on the basis of the strength of the forces that will be needed to stop the cars in the same distance. That is, put first the car on which the strongest force will have to be applied to stop it in x meters, and put last the car on which the weakest force will be applied to stop the car in the same distance.

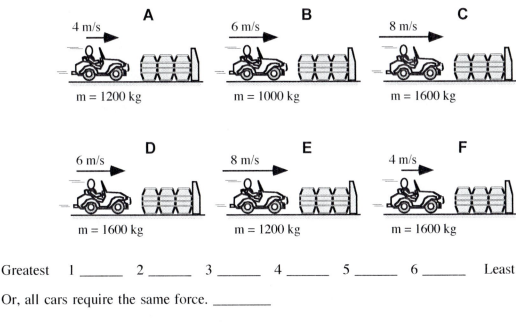

A	B	C
4 m/s	6 m/s	8 m/s
m = 1200 kg	m = 1000 kg	m = 1600 kg

D	E	F
6 m/s	8 m/s	4 m/s
m = 1600 kg	m = 1200 kg	m = 1600 kg

Greatest   1 _____   2 _____   3 _____   4 _____   5 _____   6 _____   Least

Or, all cars require the same force. _____

Please carefully explain your reasoning.

How sure were you of your ranking? (circle one)

Basically Guessed				Sure				Very Sure	
1	2	3	4	5	6	7	8	9	10

[57] T. O'Kuma

Cars and Barriers—Stopping Distance with the Same Force[58]

Shown below are six cars that are moving along horizontal roads at specified speeds. Also given are the masses of the cars. All of the cars are the same size and shape, but they are carrying loads with different masses. All of these cars are going to be stopped by plowing into identical barriers. All of the cars are going to be stopped by the same constant force by the barrier.

Rank these situations from longest to shortest on the basis of the stopping distance that will be needed to stop the cars with the same force. That is, put first the car that requires the longest stopping distance and put last the car that requires the shortest distance to stop the car with the same force.

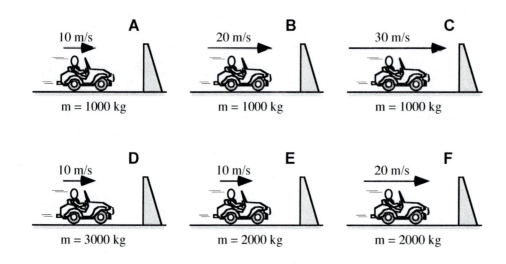

Longest   1 _____   2 _____   3 _____   4 _____   5 _____   6 _____   Shortest

Or, all the cars require the same distance. _____

Please carefully explain your reasoning.

How sure were you of your ranking? (circle one)
Basically Guessed   Sure   Very Sure
1   2   3   4   5   6   7   8   9   10

---
[58] T. O'Kuma, D. Maloney, C. Hieggelke
Ranking Task Exercises in Physics: Student Edition   58

Cars—Work Done in Change of Velocity[59]

The six situations below show *before* and *after* "snapshots" of a car's velocity. All cars have the same mass.

Rank these situations, in terms of work done on the car, from most positive to most negative, to create these changes in velocity for the same distance traveled. All cars have the same mass. Negative numbers, if any, rank lower than positive ones (-20 m/s < -10 m/s < 0 < 5 m/s).

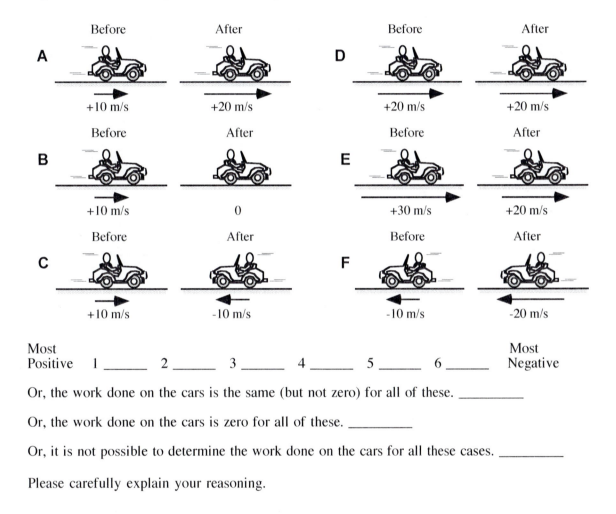

Most
Positive   1 _____ 2 _____ 3 _____ 4 _____ 5 _____ 6 _____   Most
Negative

Or, the work done on the cars is the same (but not zero) for all of these. _____

Or, the work done on the cars is zero for all of these. _____

Or, it is not possible to determine the work done on the cars for all these cases. _____

Please carefully explain your reasoning.

How sure were you of your ranking? (circle one)
Basically Guessed   Sure   Very Sure
1   2   3   4   5   6   7   8   9   10

[59] J. Cole, D. Maloney
Ranking Task Exercises in Physics: Student Edition   59

Bouncing Cart—Change in Kinetic Energy[60]

A cart with a spring plunger runs into a fixed barrier. The mass of the cart, its velocity just before impact with the barrier, and its velocity right after collision are given in each figure.

Rank the change in kinetic energy for each cart from the greatest change in kinetic energy to the least change in kinetic energy (+ direction is to the right and - to the left, with - 4 < -2).

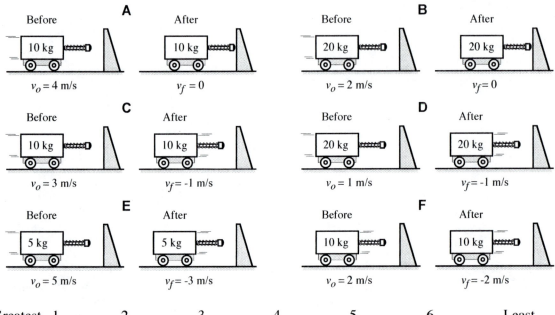

Greatest 1_____ 2_____ 3_____ 4_____ 5_____ 6_____ Least

Or, all the changes in kinetic energy are the same._____

Or, there is no change in kinetic energy for all these cases._____

Please carefully explain your reasoning.

How sure were you of your ranking? (circle one)

Basically Guessed   Sure   Very Sure
1   2   3   4   5   6   7   8   9   10
_____

[60] T. O'Kuma, C. Hieggelke

Bouncing Cart—Work Done by the Barrier[61]

A cart with a spring plunger runs into a fixed barrier. The mass of the cart, its velocity just before impact with the barrier, and its velocity right after collision are given in each figure.

Rank the work done by the barrier on each cart from the greatest work done by the barrier to the least work done by the barrier (+ direction is to the right and - to the left with - 4 < -2).

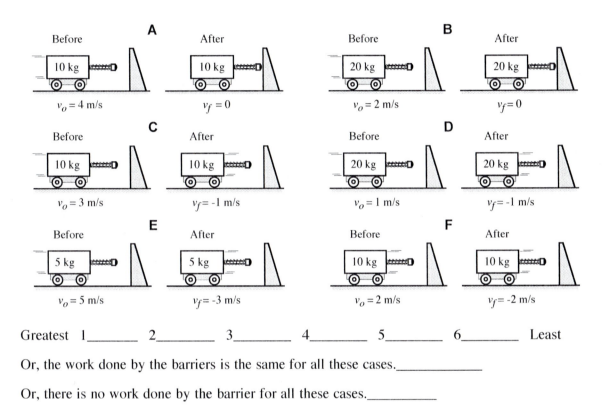

Greatest 1_____ 2_____ 3_____ 4_____ 5_____ 6_____ Least

Or, the work done by the barriers is the same for all these cases._____

Or, there is no work done by the barrier for all these cases._____

Please carefully explain your reasoning.

How sure were you of your ranking? (circle one)

Basically Guessed   Sure   Very Sure

1   2   3   4   5   6   7   8   9   10

[61] T. O'Kuma, C. Hieggelke

Bouncing Cart—Work Done on the Barrier[62]

A cart with a spring plunger runs into a fixed barrier. The mass of the cart, its velocity just before impact with the barrier, and its velocity right after collision are given in each figure.

Rank the work done on the barrier by each cart from the greatest work done on the barrier to the least work done on the barrier (+ direction is to the right and - to the left with - 4 < -2).

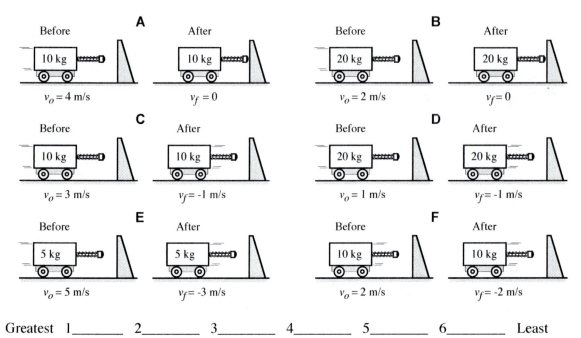

A	
Before	After
10 kg	10 kg
$v_o = 4$ m/s	$v_f = 0$

B	
Before	After
20 kg	20 kg
$v_o = 2$ m/s	$v_f = 0$

C	
Before	After
10 kg	10 kg
$v_o = 3$ m/s	$v_f = -1$ m/s

D	
Before	After
20 kg	20 kg
$v_o = 1$ m/s	$v_f = -1$ m/s

E	
Before	After
5 kg	5 kg
$v_o = 5$ m/s	$v_f = -3$ m/s

F	
Before	After
10 kg	10 kg
$v_o = 2$ m/s	$v_f = -2$ m/s

Greatest 1_____ 2_____ 3_____ 4_____ 5_____ 6_____ Least

Or, the work done on the barriers by the carts is the same for all these cases._____

Or, there is no work done on the barrier by the carts for all these cases._____

Please carefully explain your reasoning.

How sure were you of your ranking? (circle one)

Basically Guessed   Sure   Very Sure

1   2   3   4   5   6   7   8   9   10

Model Rockets—Kinetic Energy[63]

The six figures below depict six model rockets that have just had their engines turned off. All of the rockets are aimed straight up, but their speeds differ. All of the rockets are the same size and shape, but they carry different loads, so their masses differ. The mass and speed for each rocket is given in each figure. (In this situation we are going to ignore any effect air resistance may have on the rockets.) At the instant when the engines are turned off, the rockets are all at the same height.

Rank these model rockets from greatest to least on the basis of the kinetic energy they have at the top of their flights.

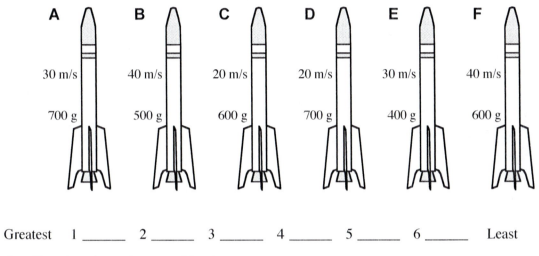

A	B	C	D	E	F
30 m/s	40 m/s	20 m/s	20 m/s	30 m/s	40 m/s
700 g	500 g	600 g	700 g	400 g	600 g

Greatest   1 _____   2 _____   3 _____   4 _____   5 _____   6 _____   Least

Or, all rockets have the same kinetic energy._____

Please carefully explain your reasoning.

How sure were you of your ranking? (circle one)

Basically Guessed				Sure				Very Sure	
1	2	3	4	5	6	7	8	9	10

[63] T. O'Kuma, D. Maloney

Sliding Masses on Incline—Kinetic Energy[64]

For the figures below, all surfaces are frictionless. All masses start from rest.

Rank in order from greatest to least the final kinetic energies of the sliding masses the instant before they reach the bottom of the incline.

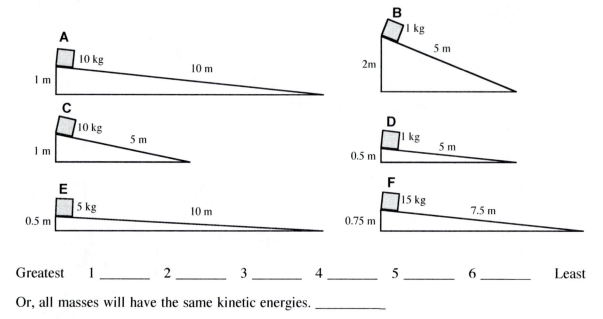

Greatest   1 _____   2 _____   3 _____   4 _____   5 _____   6 _____   Least

Or, all masses will have the same kinetic energies. _____

Please carefully explain your reasoning.

How sure were you of your ranking? (circle one)
Basically Guessed   Sure   Very Sure
1   2   3   4   5   6   7   8   9   10

Sliding Masses on Incline—Change in Potential Energy[65]

For the figures below, all surfaces are frictionless. All masses start from rest at the top of the incline.

Rank in order from greatest to least the change in gravitational potential energy of the sliding masses from the top of the incline to the bottom of the incline.

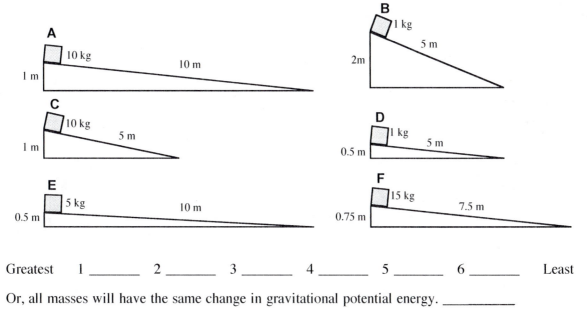

Greatest   1 _____ 2 _____ 3 _____ 4 _____ 5 _____ 6 _____   Least

Or, all masses will have the same change in gravitational potential energy. _____

Please carefully explain your reasoning.

How sure were you of your ranking? (circle one)
Basically Guessed   Sure   Very Sure
1   2   3   4   5   6   7   8   9   10

[65] P. Marquard, D. Maloney

Cars—Change in Kinetic Energy During a Change of Velocity[66]

The six situations below show *before* and *after* "snapshots" of a car's velocity. All cars have the same mass and have traveled the same distance during this change.

Rank these situations, in terms of the change in kinetic energy of these cars, from most positive to most negative. Negative numbers, if any, rank lower than positive ones (-20 m/s < -10 m/s < 0 < 5 m/s).

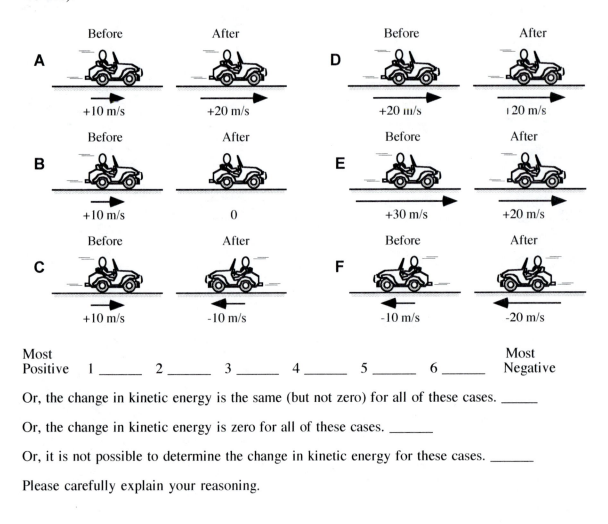

Most
Positive   1 _____   2 _____   3 _____   4 _____   5 _____   6 _____   Most
Negative

Or, the change in kinetic energy is the same (but not zero) for all of these cases. _____

Or, the change in kinetic energy is zero for all of these cases. _____

Or, it is not possible to determine the change in kinetic energy for these cases. _____

Please carefully explain your reasoning.

How sure were you of your ranking? (circle one)
Basically Guessed   Sure   Very Sure
1   2   3   4   5   6   7   8   9   10

[66] J. Cole, D. Maloney, C. Hieggelke

Ball Motion Diagram—Kinetic Energy[67]

The following drawings indicate the motion of a ball subject to one or more forces on various surfaces from left to right. Each circle represents the position of the ball at succeeding instants of time. Each time-interval between positions is equal. In all situations, the balls start with the same velocity.

Rank each case from the highest to the lowest final kinetic energy based on the figures using the coordinate system shown in the diagram. Assume the acceleration for each situation to be constant.

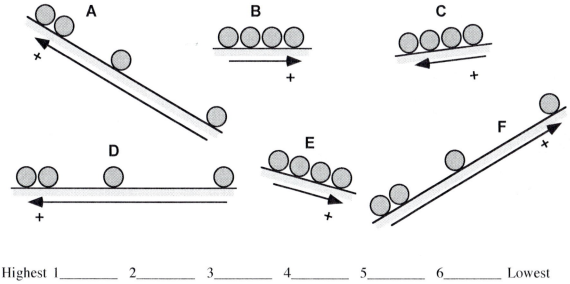

Highest 1_____ 2_____ 3_____ 4_____ 5_____ 6_____ Lowest

Or, all have the same final kinetic energy._____

Please carefully explain your reasoning.

How sure were you of your ranking? (circle one)

Basically Guessed				Sure				Very Sure	
1	2	3	4	5	6	7	8	9	10

[67] D. Maloney
Ranking Task Exercises in Physics: Student Edition   67

Equal Forces on Boxes—Work Done on Box[68]

In the figures below, identical boxes of mass 10 kg are moving at the same initial velocity to the right on a flat surface. The same magnitude force, F, is applied to each box for the distance, d, indicated in the figures.

Rank these situations from greatest to least in order of the work done on the box by F while the box moves the indicated distance to the right.

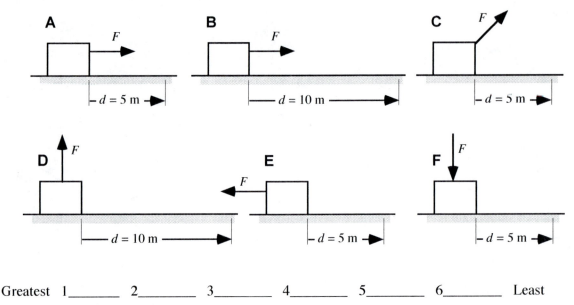

Greatest 1_____ 2_____ 3_____ 4_____ 5_____ 6_____ Least

Or, all of the boxes experience the same work. _____

Please carefully explain your reasoning.

How sure were you of your ranking? (circle one)

Basically Guessed   Sure   Very Sure
1   2   3   4   5   6   7   8   9   10

Equal Force on Boxes—Work Done on Hand[69]

In the figures below, identical boxes of mass 10 kg are moving at the same initial velocity to the right on a flat surface. The same magnitude force, F, is applied by a hand to each box for the distance, d, indicated in the figures.

Rank these situations from greatest to least in order of the work done by the box on the hand causing F while the box moves the indicated distance to the right.

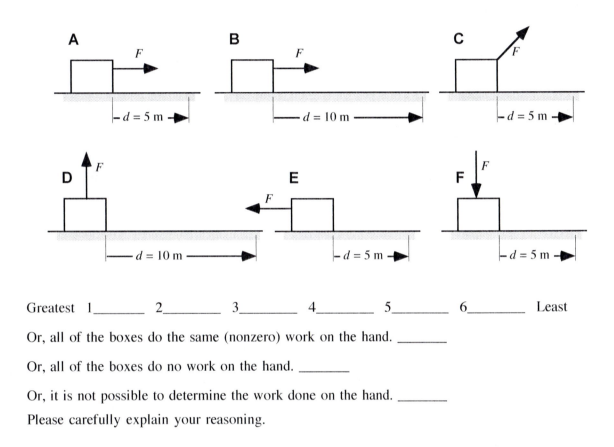

Greatest 1_____ 2_____ 3_____ 4_____ 5_____ 6_____ Least

Or, all of the boxes do the same (nonzero) work on the hand. _____

Or, all of the boxes do no work on the hand. _____

Or, it is not possible to determine the work done on the hand. _____

Please carefully explain your reasoning.

How sure were you of your ranking? (circle one)
Basically Guessed   Sure   Very Sure
1   2   3   4   5   6   7   8   9   10

[69] E. Eckard, D. Maloney, C. Hieggelke

Velocity Time Graph—Work Done on Box[70]

Shown below is a graph of velocity versus time for an object that moves along a straight, horizontal line under the, perhaps intermittent, action of a single force exerted by an external agent.

Rank the intervals shown on the graph, from greatest to least, on the basis of the work done on the object by the external agent.

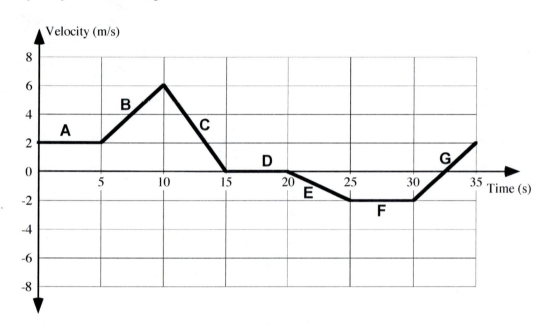

Greatest   1_____ 2_____ 3_____ 4_____ 5_____ 6_____ 7 _____   Least

Or, the agent does the same amount of work during all of these intervals. _____

Or, the agent does not work during any of these intervals. _____

Please carefully explain your reasoning.

How sure were you of your ranking? (circle one)

Basically Guessed				Sure				Very Sure	
1	2	3	4	5	6	7	8	9	10

Pendulums—Maximum Speed of the Bob[71]

Shown below are six situations where spheres are attached to strings forming pendulums. The pendulums vary in mass and length, but the angles from the vertical are the same for all.

Rank these situations, from greatest to least, on the basis of the maximum speed of the bob at the bottom of the swing. In other words, put first the pendulum whose bob has the greatest speed going through the equilibrium point and put last the pendulum whose bob has the least speed at equilibrium.

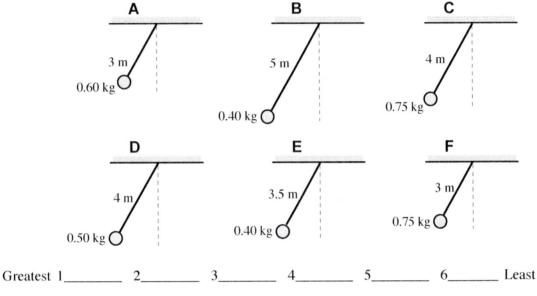

Greatest 1_____ 2_____ 3_____ 4_____ 5_____ 6_____ Least

Or, the maximum speed is the same for all six of these. _____

Please carefully explain your reasoning.

How sure were you of your ranking? (circle one)

Basically Guessed				Sure				Very Sure	
1	2	3	4	5	6	7	8	9	10

[71] C. Hieggelke, D. Maloney, T. O'Kuma
Ranking Task Exercises in Physics: Student Edition   71

Force Pushing Box—Change in Kinetic Energy[72]

Various similar boxes are being pushed for 10 m across a floor by a *net* horizontal force as shown below. The mass of the boxes and the net horizontal force for each case are given in the indicated figures.

Rank the change in kinetic energy for each box from the greatest change in kinetic energy to the least change in kinetic energy. All boxes have an initial velocity of +10 m/s (+ direction is to the right and - to the left with - 4 < -2).

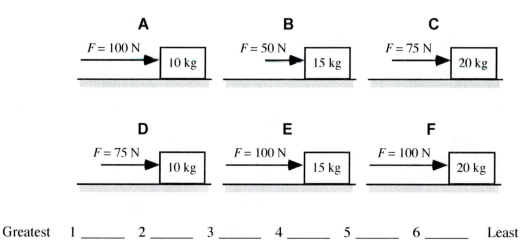

Greatest 1 _____ 2 _____ 3 _____ 4 _____ 5 _____ 6 _____ Least

Or, all changes in kinetic energy are the same._____

Please carefully explain your reasoning.

How sure were you of your ranking? (circle one)

Basically Guessed   Sure   Very Sure

1   2   3   4   5   6   7   8   9   10

Pushing Box with Friction—Change in Kinetic Energy[73]

Various similar boxes are being pushed for 10 m across a floor by a horizontal force as shown below. The weights of the boxes and the applied horizontal force for each case are given in the indicated figures. The frictional force is 20% of the weight of the box ($g = 10$ N/kg).

Rank the change in kinetic energy for each box from the greatest change in kinetic energy to the least change in kinetic energy. All boxes have an initial velocity of +10 m/s (+ direction is to the right and - to the left, with - 4 < -2).

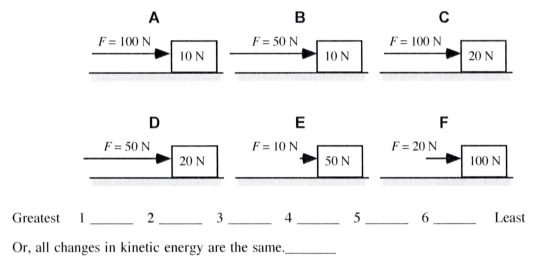

Greatest   1 _____   2 _____   3 _____   4 _____   5 _____   6 _____   Least

Or, all changes in kinetic energy are the same._____

Please carefully explain your reasoning.

How sure were you of your ranking? (circle one)

Basically Guessed				Sure				Very Sure	
1	2	3	4	5	6	7	8	9	10

[73] C. Hieggelke, T. O'Kuma

Cars and Barriers—Stopping Time with the Same Force[74]

Shown below are six cars that are moving along horizontal roads at specified speeds. Also given are the masses of the cars. All of the cars are the same size and shape, but they are carrying loads with different masses. All of these cars are going to be stopped by plowing into identical barriers. All of the cars are going to be stopped by the same constant force by the barrier.

Rank these situations from longest to shortest on the basis of the stopping time that will be needed to stop the cars with the same force. That is, put first the car that requires the longest time and put last the car that requires the shortest time to stop the car with the same force.

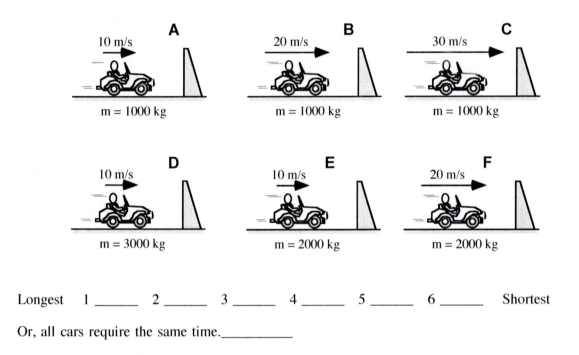

Longest   1 _____   2 _____   3 _____   4 _____   5 _____   6 _____   Shortest

Or, all cars require the same time._____

Please carefully explain your reasoning.

How sure were you of your ranking? (circle one)
Basically Guessed   Sure   Very Sure
1   2   3   4   5   6   7   8   9   10

[74] T. O'Kuma, C. Hieggelke, D. Maloney
Ranking Task Exercises in Physics: Student Edition   74

Exploding Shells—Final Location of Center of Mass[75]

Shown below are six situations where shells are at the top of their trajectories. These shells explode, at this instant, into two pieces. One piece, m_1, falls straight down (vertically) to the ground. All of these projectiles were fired from the same point at the same angle. We are told the speed of the shell at the top of the trajectory and the masses of the two pieces. Ignore air resistance in this situation.

Rank these situations, from greatest to least, on the basis of how far from the launch point the center of mass of the projectile ends up. That is, put first the situation where the final location of the center of mass is farthest from the launch point and put last the situation where it ends up closest to the launch point.

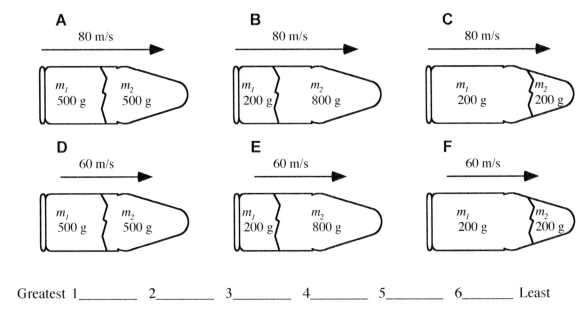

Greatest 1_____ 2_____ 3_____ 4_____ 5_____ 6_____ Least

Or, all six of the centers of mass end up the same distance from the launch point. _____

Please carefully explain your reasoning.

How sure were you of your ranking? (circle one)

Basically Guessed				Sure				Very Sure	
1	2	3	4	5	6	7	8	9	10

[75] D. Maloney

Bouncing Cart—Change in Momentum I[76]

A cart with a spring plunger runs into a fixed barrier. The mass of the cart, its velocity just before impact with the barrier, and its velocity right after collision are given in each figure.

Rank the change in momentum for each cart from the greatest change in momentum to the least change in momentum (+ direction is to the right and - to the left with - 4 < -2).

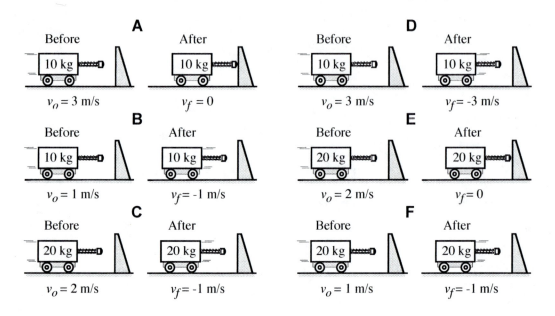

A

Before — 10 kg — $v_o = 3$ m/s
After — 10 kg — $v_f = 0$

B

Before — 10 kg — $v_o = 1$ m/s
After — 10 kg — $v_f = -1$ m/s

C

Before — 20 kg — $v_o = 2$ m/s
After — 20 kg — $v_f = -1$ m/s

D

Before — 10 kg — $v_o = 3$ m/s
After — 10 kg — $v_f = -3$ m/s

E

Before — 20 kg — $v_o = 2$ m/s
After — 20 kg — $v_f = 0$

F

Before — 20 kg — $v_o = 1$ m/s
After — 20 kg — $v_f = -1$ m/s

Greatest 1_____ 2_____ 3_____ 4_____ 5_____ 6_____ Least

Or, all the changes in momentum are the same._____

Please carefully explain your reasoning.

How sure were you of your ranking? (circle one)

Basically Guessed   Sure   Very Sure

1   2   3   4   5   6   7   8   9   10

Bouncing Cart—Change in Momentum II[77]

A cart with a spring plunger runs into a fixed barrier. The mass of the cart, its velocity just before impact with the barrier, and its velocity right after collision are given in each figure.

Rank the change in momentum for each cart from the greatest change in momentum to the least change in momentum (+ direction is to the right and - to the left with - 4 < -2).

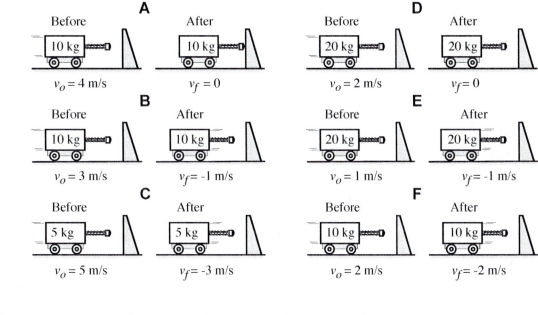

A

Before — 10 kg — $v_o = 4$ m/s
After — 10 kg — $v_f = 0$

B

Before — 10 kg — $v_o = 3$ m/s
After — 10 kg — $v_f = -1$ m/s

C

Before — 5 kg — $v_o = 5$ m/s
After — 5 kg — $v_f = -3$ m/s

D

Before — 20 kg — $v_o = 2$ m/s
After — 20 kg — $v_f = 0$

E

Before — 20 kg — $v_o = 1$ m/s
After — 20 kg — $v_f = -1$ m/s

F

Before — 10 kg — $v_o = 2$ m/s
After — 10 kg — $v_f = -2$ m/s

Greatest 1_____ 2_____ 3_____ 4_____ 5_____ 6_____ Least

Or, all the changes in momentum are the same._____

Please carefully explain your reasoning.

How sure were you of your ranking? (circle one)

Basically Guessed				Sure				Very Sure	
1	2	3	4	5	6	7	8	9	10

[77] T. O'Kuma

Bouncing Cart—Change in Momentum III[78]

A cart with a spring plunger runs into a fixed barrier. The mass of the cart, its velocity just before impact with the barrier, and its velocity right after collision are given in each figure.

Rank the change in momentum for each cart from the greatest change in momentum to the least change in momentum (+ direction is to the right and - to the left with - 4 < -2).

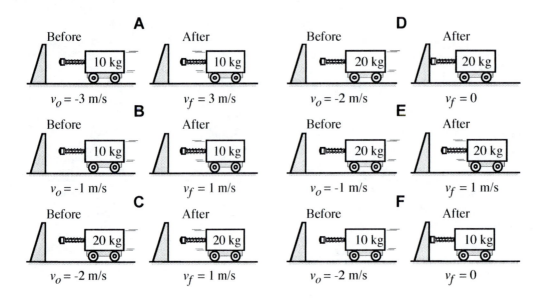

A
Before — 10 kg — $v_0 = -3$ m/s
After — 10 kg — $v_f = 3$ m/s

B
Before — 10 kg — $v_0 = -1$ m/s
After — 10 kg — $v_f = 1$ m/s

C
Before — 20 kg — $v_0 = -2$ m/s
After — 20 kg — $v_f = 1$ m/s

D
Before — 20 kg — $v_0 = -2$ m/s
After — 20 kg — $v_f = 0$

E
Before — 20 kg — $v_0 = -1$ m/s
After — 20 kg — $v_f = 1$ m/s

F
Before — 10 kg — $v_0 = -2$ m/s
After — 10 kg — $v_f = 0$

Greatest 1_____ 2_____ 3_____ 4_____ 5_____ 6_____ Least

Or, all the changes in momentum are the same._____

Please carefully explain your reasoning.

How sure were you of your ranking? (circle one)

Basically Guessed   Sure   Very Sure

1   2   3   4   5   6   7   8   9   10

Force Pushing Box—Change in Momentum[79]

Various similar boxes are being pushed for 10 seconds across a floor by a net horizontal force as shown below. The mass of the boxes and the net horizontal force for each case are given in the indicated figures.

Rank the change in momentum for each box from the greatest change in momentum to the least change in momentum. All boxes have an initial velocity of 0 m/s (+ direction is to the right and - to the left with - 4 < -2).

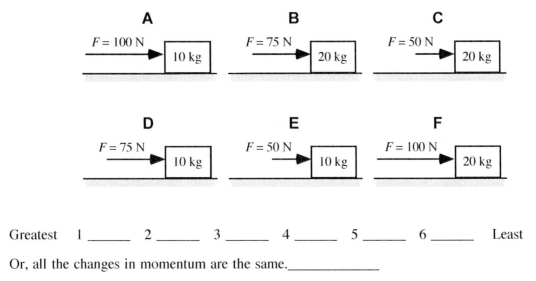

Greatest   1 _____   2 _____   3 _____   4 _____   5 _____   6 _____   Least

Or, all the changes in momentum are the same. _____

Please carefully explain your reasoning.

How sure were you of your ranking? (circle one)

Basically Guessed   Sure   Very Sure

1   2   3   4   5   6   7   8   9   10

Force Pushing Box—Final Momentum[80]

Various similar boxes are being pushed for 10 seconds across a floor by a net horizontal force as shown below. The mass of the boxes and the net horizontal force for each case is given in the indicated figures.

Rank the final momentum for each box from the greatest momentum to the least momentum. All boxes have an initial velocity of 10 m/s. Motion to the right is positive.

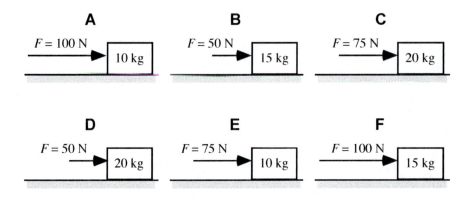

Greatest 1_____ 2_____ 3_____ 4_____ 5_____ 6_____ Least

Or, all the final momenta are the same._____

Please carefully explain your reasoning.

How sure were you of your ranking? (circle one)
Basically Guessed   Sure   Very Sure
1   2   3   4   5   6   7   8   9   10

[80] T. O'Kuma

Ranking Task Exercises in Physics: Student Edition   80

Cars—Impulse During a Change of Velocity[81]

The six situations below show *before* and *after* "snapshots" of a car's velocity. All cars have the same mass.

Rank these situations, in terms of impulse on these cars, from most positive to most negative, to create these changes in velocity. Negative numbers, if any, rank lower than positive ones (-20 m/s < -10 m/s < 0 < 5 m/s).

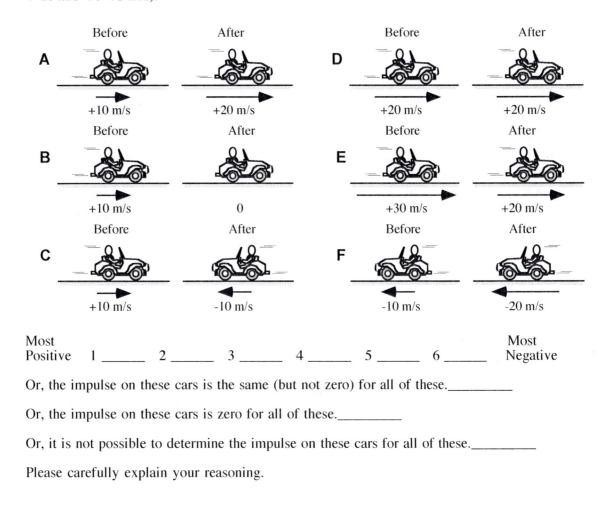

Most
Positive   1 _____ 2 _____ 3 _____ 4 _____ 5 _____ 6 _____   Most
Negative

Or, the impulse on these cars is the same (but not zero) for all of these._____

Or, the impulse on these cars is zero for all of these._____

Or, it is not possible to determine the impulse on these cars for all of these._____

Please carefully explain your reasoning.

How sure were you of your ranking? (circle one)
Basically Guessed				Sure				Very Sure	
1	2	3	4	5	6	7	8	9	10

Cars—Change in Momentum During a Change of Velocity[82]

The six situations below show *before* and *after* "snapshots" of a car's velocity. All cars have the same mass.

Rank these situations, in terms of the change in momentum of these cars, from most positive to most negative. Note that negative numbers, if any, rank lower than positive ones (-20 m/s < -10 m/s < 0 < 5 m/s).

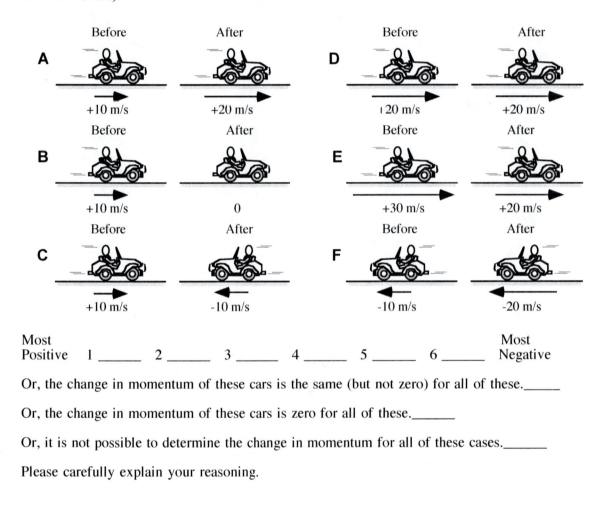

Before			**After**		**Before**		**After**

A   Before +10 m/s   After +20 m/s   D   Before 20 m/s   After +20 m/s

B   Before +10 m/s   After 0   E   Before +30 m/s   After +20 m/s

C   Before +10 m/s   After -10 m/s   F   Before -10 m/s   After -20 m/s

Most
Positive   1 _____   2 _____   3 _____   4 _____   5 _____   6 _____   Most Negative

Or, the change in momentum of these cars is the same (but not zero) for all of these._____

Or, the change in momentum of these cars is zero for all of these._____

Or, it is not possible to determine the change in momentum for all of these cases._____

Please carefully explain your reasoning.

How sure were you of your ranking? (circle one)
Basically Guessed   Sure   Very Sure
1   2   3   4   5   6   7   8   9   10

Four L's Rotating About an Axis (side view)—Moment of Inertia[83]

Below are four identical figure **L**'s, which are constructed from two uniform rods of equal lengths and masses. For each figure, a different axis of rotation is indicated by the small circle with the dot inside, which indicates an axis that is perpendicular to the plane of the **L**'s. The axis of rotation is located either at the center or one end of a rod for each figure.

Rank these **L** figures according to their moments of inertia about the indicated axes, from largest to smallest. Ignore the width of each rod but not the length.

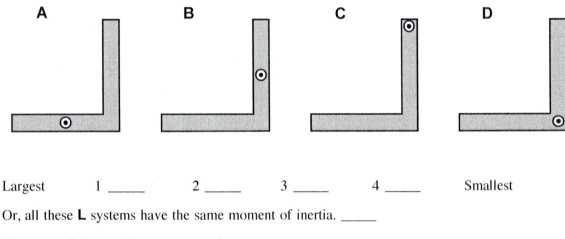

Largest   1 ____   2 ____   3 ____   4 ____   Smallest

Or, all these **L** systems have the same moment of inertia. ____

Please carefully explain your reasoning.

How sure were you of your ranking? (circle one)

Basically Guessed				Sure					Very Sure
1	2	3	4	5	6	7	8	9	10

[83] C. Hieggelke

Five T's Rotating About an Axis (top view)—Moment of Inertia[84]

Below are five identical figure **T**'s, which are constructed from two uniform rods of equal lengths and masses. For each figure, a different axis of rotation in the plane of the paper is indicated by the dotted line. The axis of rotation is located either at the center or one end of a rod for each figure.

Rank these **T** figures according to their moments of inertia about the indicated axes, from largest to smallest. Ignore the width of each rod but not the length.

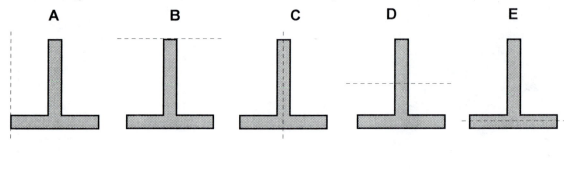

Largest   1 _____   2 _____   3 _____   4 _____   5 _____   Smallest

Or, all these **T** systems have the same moment of inertia. _____

Please carefully explain your reasoning.

How sure were you of your ranking? (circle one)

Basically Guessed				Sure					Very Sure	
1	2	3	4	5	6	7	8	9	10	

[84] C. Hieggelke

Four T's Rotating About an Axis (side view)—Moment of Inertia[85]

Below are four identical figure **T**'s, which are constructed from two uniform rods of equal lengths and masses. For each figure, a different axis of rotation is indicated by the small circle with the dot inside, which indicates an axis that is perpendicular to the plane of the **T**'s. The axis of rotation is located either at the center or one end of a rod for each figure.

Rank these **T** figures according to their moments of inertia about the indicated axes, from largest to smallest. Ignore the width of each rod but not the length.

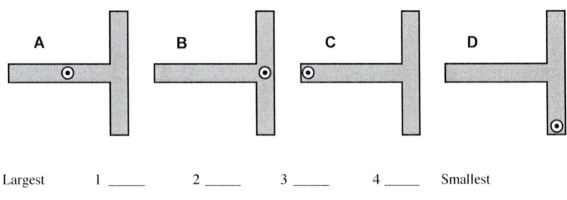

Largest   1 _____   2 _____   3 _____   4 _____   Smallest

Or, all these **T** systems have the same moment of inertia._____

Please carefully explain your reasoning.

How sure were you of your ranking? (circle one)

Basically Guessed				Sure				Very Sure	
1	2	3	4	5	6	7	8	9	10

[85] C. Hieggelke

Five T's Rotating About an Axis (top view)—Net Gravitational Torque[86]

Shown below (in a top view with the gravitation force into the page) are five identical figure **T**'s, which are constructed from two uniform rods of equal lengths and masses. For each figure, a different axis of rotation, which is in the plane of the page, is indicated by the dotted line. The axes are either at the center or one end of a rod.

Rank these **T** figures according to the magnitudes of the net torque about the indicated axes, from largest to smallest.

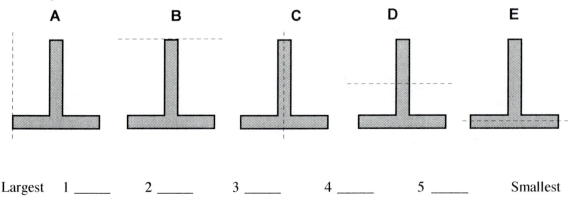

Largest   1 _____   2 _____   3 _____   4 _____   5 _____   Smallest

Or, all these **T** systems have the same magnitude of the net torque. _____

Please carefully explain your reasoning.

How sure were you of your ranking? (circle one)
Basically Guessed   Sure   Very Sure
1   2   3   4   5   6   7   8   9   10

Five T's Rotating About an Axis (top view)—Angular Acceleration[87]

Shown below (in a top view with the gravitation force into the page ⊕) are five identical figure **T**'s, which are constructed from two uniform rods of equal lengths and masses. For each figure, a different axis of rotation, which is in the plane of the page, is indicated by the dotted line. The axes are either at the center or one end of a rod.

Rank these **T** figures according to the <u>magnitude</u> of the initial angular acceleration about the indicated axes, from largest to smallest. Ignore the width of each rod but not the length.

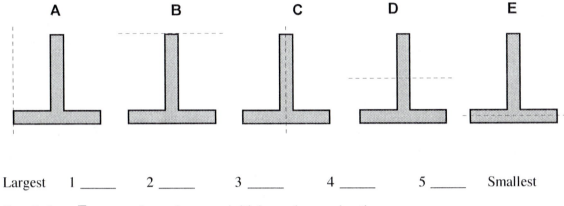

Largest   1 _____   2 _____   3 _____   4 _____   5 _____   Smallest

Or, all these **T** systems have the same initial angular acceleration. _____

Please carefully explain your reasoning.

How sure were you of your ranking? (circle one)
Basically Guessed   Sure   Very Sure
1   2   3   4   5   6   7   8   9   10

_____

Blocks on Rotating Turntables—Horizontal Force[88]

Shown below in a top view are six blocks that are sitting at rest on rotating turntables. All of the turntables have the same rotation rate. The masses of the blocks and how far out from the center they sit varies. Specific values of the variable are given in the figures.

Rank these blocks, from greatest to least, on the basis of the magnitude of the horizontal forces holding the blocks on the turntables. That is, put first the block that has the largest force holding it on the turntable and put last the block that has the weakest force holding it on the turntable.

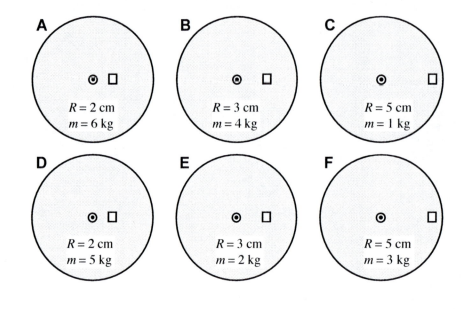

A
$R = 2$ cm
$m = 6$ kg

B
$R = 3$ cm
$m = 4$ kg

C
$R = 5$ cm
$m = 1$ kg

D
$R = 2$ cm
$m = 5$ kg

E
$R = 3$ cm
$m = 2$ kg

F
$R = 5$ cm
$m = 3$ kg

Greatest 1_____ 2_____ 3_____ 4_____ 5_____ 6_____ Least

Or, all of these blocks are held by equal strength forces. _____

Please carefully explain your reasoning.

How sure were you of your ranking? (circle one)

Basically Guessed   Sure   Very Sure

1   2   3   4   5   6   7   8   9   10

[88] C. Hieggelke, D. Maloney, T. O'Kuma

Ranking Task Exercises in Physics: Student Edition   88

Hanging Weights and Fixed Disks—Torque[89]

Shown below are six situations where vertically oriented circular disks have strings wrapped around them. The other ends of the strings are attached to hanging masses. The radii of the disks, the masses of the disks, and the masses of the hanging masses all vary. The disks are fixed and are *not* free to rotate. Specific values of the variables are given in the figures.

Rank these situations, from greatest to least, on the basis of the magnitude of the torque about the center of the disks. That is, put first the situation where the disk has the greatest torque acting on it and put last the situation where the disk has the least torque acting on it.

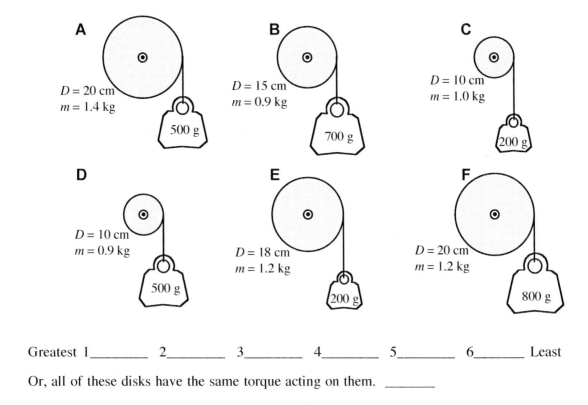

A $D = 20$ cm $m = 1.4$ kg — 500 g

B $D = 15$ cm $m = 0.9$ kg — 700 g

C $D = 10$ cm $m = 1.0$ kg — 200 g

D $D = 10$ cm $m = 0.9$ kg — 500 g

E $D = 18$ cm $m = 1.2$ kg — 200 g

F $D = 20$ cm $m = 1.2$ kg — 800 g

Greatest 1_____ 2_____ 3_____ 4_____ 5_____ 6_____ Least

Or, all of these disks have the same torque acting on them. _____

Please carefully explain your reasoning.

How sure were you of your ranking? (circle one)
Basically Guessed   Sure   Very Sure
1   2   3   4   5   6   7   8   9   10

[89] C. Hieggelke, D. Maloney, T. O'Kuma
Ranking Task Exercises in Physics: Student Edition   89

Horizontal Uniform Rods—Angular Acceleration[90]

Shown below in a top view are six uniform rods that vary in mass (*M*) and length (*L*). Also shown are circles representing a vertical axis around which the rods are going to be rotated in a horizontal plane and arrows representing forces acting to rotate the rods. The forces change direction in order to always act perpendicular to the rods. Specific values for the lengths and masses of the rods and the magnitudes of the forces are given in each figure.

Rank these rods, from greatest to least, on the basis of the magnitude of their angular acceleration. That is, put first the rod that has the largest angular acceleration and put last the one that will have the smallest angular acceleration.

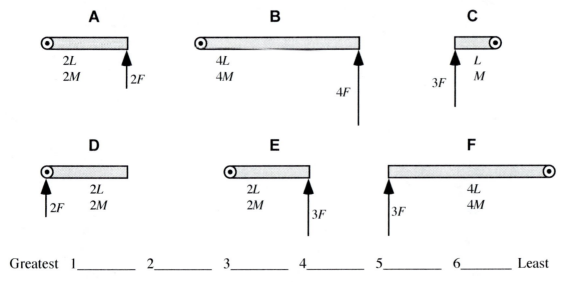

Greatest 1_____ 2_____ 3_____ 4_____ 5_____ 6_____ Least

Or, all six of these rods will have the same magnitude angular acceleration. _____

Please carefully explain your reasoning.

How sure were you of your ranking? (circle one)
Basically Guessed   Sure   Very Sure
1   2   3   4   5   6   7   8   9   10

---

[90] C. Hieggelke, D. Maloney, T. O'Kuma
Ranking Task Exercises in Physics: Student Edition   90

Horizontal Uniform Rods—Change in Angular Momentum[91]

Shown below in a top view are six uniform rods that vary in mass (M) and length (L). Also shown are circles representing a vertical axis around which the rods are going to be rotated in a horizontal plane and arrows representing forces acting to rotate the rods. The forces change direction in order to always act perpendicular to the rods. Specific values for the lengths and masses of the rods and the magnitudes of the forces are given in each figure.

Rank these rods, from greatest to least, on the basis of their change in the magnitude of angular momentum for the same time period. That is, put first the rod that has the largest change in angular momentum and put last the one that will have the smallest change.

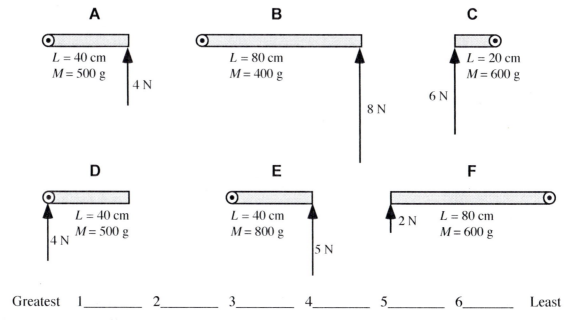

Greatest   1_____ 2_____ 3_____ 4_____ 5_____ 6_____ Least

Or, all six of these rods will have the same change in magnitude of angular momentum. _____

Please carefully explain your reasoning.

How sure were you of your ranking? (circle one)
Basically Guessed   Sure   Very Sure
1   2   3   4   5   6   7   8   9   10

[91] C. Hieggelke, D. Maloney, T. O'Kuma
Ranking Task Exercises in Physics: Student Edition   91

Rotating Systems of Point Masses—Difficult to Rotate[92]

Shown below are six arrangements of 10-point masses. Each of the point masses is the same size and has the same mass. Also shown in each figure is a solid line representing an axis about which the masses are going to be rotated. The point masses exert forces on each other so that they all maintain the arrangements shown while being rotated.

Rank these arrangements, from greatest to least, on the basis of how hard it will be to start the arrangements rotating. That is, put first the arrangement that will be the most difficult to start rotating and put last the easiest arrangement.

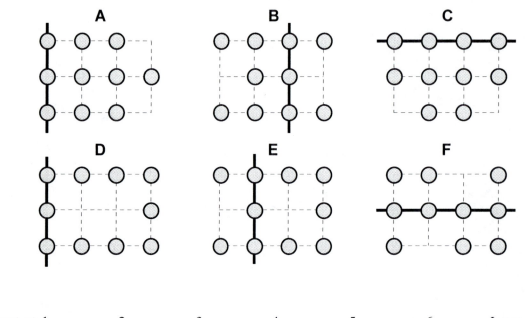

Greatest 1_____ 2_____ 3_____ 4_____ 5_____ 6_____ Least

Or, all of these arrangements will be equally difficult to rotate. _____

Please carefully explain your reasoning.

How sure were you of your ranking? (circle one)

Basically Guessed					Sure				Very Sure
1	2	3	4	5	6	7	8	9	10

[92] C. Hieggelke, D. Maloney, T. O'Kuma

Rotating Systems of Point Masses—Center of Mass[93]

Shown below are six arrangements of 10 point masses. Each of the point masses is the same size and has the same mass. Also shown in each figure is a solid line representing an axis about which the masses are going to be rotated. The point masses exert forces on each other so that they all maintain the arrangements shown while being rotated.

Rank these arrangements, from greatest to least, on the basis of the distance between the center of mass and the axis of rotation.

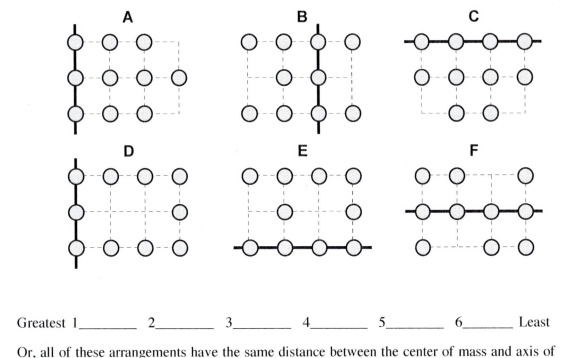

Greatest 1_____ 2_____ 3_____ 4_____ 5_____ 6_____ Least

Or, all of these arrangements have the same distance between the center of mass and axis of rotation. _____

Please carefully explain your reasoning.

How sure were you of your ranking? (circle one)

Basically Guessed				Sure				Very Sure	
1	2	3	4	5	6	7	8	9	10

Statics—Difficult to Hold I[94]

Shown below are six situations where a student is holding a meter stick at the left end at various angles. A 1000 g mass is hung on the meter sticks at different locations. All of the meter sticks are identical, but the distance along the meter stick at which the 1000 g mass is hung and the angles at which the student holds the meter stick vary. Specific values are given in each figure. (Ignore the mass of the meter stick.)

Rank these situations, from hardest to easiest, on the basis of how difficult it would be for the student to hold the meter stick from the left end in the position shown. That is, put first the situation where it would be hardest to hold the meter stick at the angle shown and put last the situation where it would be easiest to hold it at the angle shown.

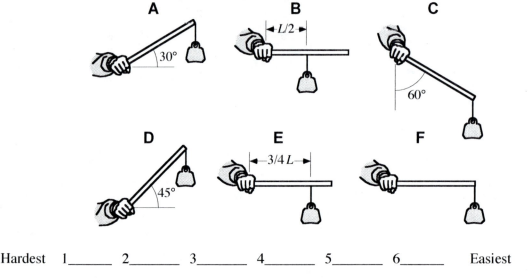

Hardest   1_____ 2_____ 3_____ 4_____ 5_____ 6_____   Easiest

Or, it would be equally difficult to hold all seven of these meter sticks as shown. _____

Please carefully explain your reasoning.

How sure were you of your ranking? (circle one)
Basically Guessed   Sure   Very Sure
1   2   3   4   5   6   7   8   9   10

_____

[94] C. Hieggelke, D. Maloney, T. O'Kuma
Ranking Task Exercises in Physics: Student Edition   94

Statics—Difficult to Hold II[95]

Shown below are six situations where a student is holding a meter stick straight out horizontally. All of the meter sticks are identical, but the number, mass, and locations (either at the 50 cm mark or at the other end of the meter stick from the student) of the objects hung on the sticks vary. The specific values and locations are given in each figure.

Rank these situations, from hardest to easiest, on the basis of how difficult it would be for the student to keep the meter stick from rotating. That is, put first the situation where it would be hardest to hold the meter stick horizontal and put last the situation where it would be easiest to hold it horizontal.

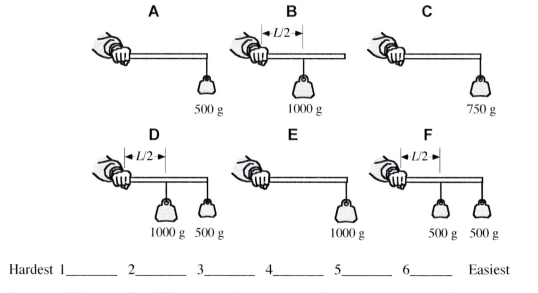

Hardest 1_____ 2_____ 3_____ 4_____ 5_____ 6_____ Easiest

Or, it would be equally difficult to hold all seven of these meter sticks as shown. _____

Please carefully explain your reasoning.

How sure were you of your ranking? (circle one)
Basically Guessed   Sure   Very Sure
1   2   3   4   5   6   7   8   9   10

---

[95] C. Hieggelke, D. Maloney, T. O'Kuma
Ranking Task Exercises in Physics: Student Edition   95

Springs and Masses—Period of Oscillating Mass[96]

Each figure below shows a block attached to the end of a spring resting on a frictionless surface. In each figure, the springs are to be stretched to the right by a distance given in the figure and released. The blocks will then proceed to oscillate. The mass is given for each different block, and the force constant is given for each different spring.

Rank the figures from greatest to least on the basis of the period of the vibratory motion. That is, rank the figures on the basis of how long each will take to go through one complete cycle.

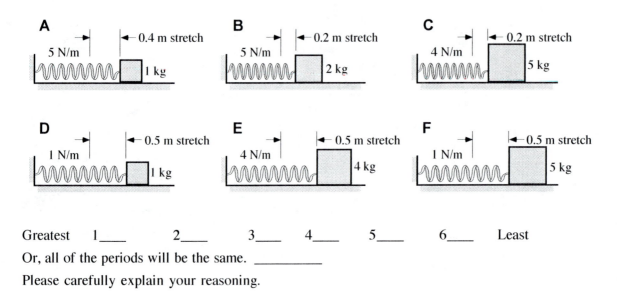

Greatest   1____   2____   3____   4____   5____   6____   Least

Or, all of the periods will be the same. _____

Please carefully explain your reasoning.

How sure were you of your ranking? (circle one)

Basically Guessed   Sure   Very Sure
1   2   3   4   5   6   7   8   9   10

SHM Equations—Period, Maximum Velocity, and Acceleration[97]

Listed below are mathematical representations for objects undergoing simple harmonic motions (SHM). In these expressions, y is in meters, t is in seconds, and the numerical values have appropriate units.

A	**B**	**C**
$y = 6 \sin 3t$	$y = 3 \sin 6t$	$y = 6 \cos 3t$
D	**E**	**F**
$y = 6 \sin (3t+30°)$	$y = 10 \cos 6t$	$y = 10 \sin 2t$

Rank these mathematical representations of SHM, from greatest to least, on the basis of the <u>period</u> or the time it takes for the object to complete one cycle of this motion.

Greatest   1_____   2_____   3_____   4_____   5_____   6_____   Least

Or, it is not possible to compare these representations without knowing more. _____

Please explain your reasoning.

Rank these mathematical representations of SHM, from greatest to least, on the basis of the maximum velocity of the object during one complete cycle of this motion.

Greatest   1_____   2_____   3_____   4_____   5_____   6_____   Least

Or, it is not possible to compare these representations without knowing more. _____

Please explain your reasoning.

Rank these mathematical representations of SHM, from greatest to least, on the basis of the maximum acceleration of the object during one complete cycle of this motion.

Greatest   1_____   2_____   3_____   4_____   5_____   6_____   Least

Or, it is not possible to compare these representations without knowing more. _____

Please explain your reasoning.

How sure were you of your ranking? (circle one)

Basically Guessed				Sure				Very Sure	
1	2	3	4	5	6	7	8	9	10

[97] C. Hieggelke, D. Maloney
Ranking Task Exercises in Physics: Student Edition   97

Blocks Suspended in Liquids—Buoyant Force[98]

Shown below are six containers that contain various liquids. Blocks of various solids are suspended in the liquids by being hung from a supporting rod. All of these blocks are the same size, but they have different masses (labeled as M_b) since they are made of different materials. All of the containers have the same volume of liquid, but the masses of these liquids vary (labeled M_l) since the liquids are different. Specific values for the masses of the blocks and the liquids are given in each figure. The volume of the blocks is one-fifth the volume of the liquids.

Rank these situations, from greatest to smallest, on the basis of the buoyant forces on the blocks. That is, put first the block that experiences the greatest buoyant force, and put last the block that experiences the smallest buoyant force.

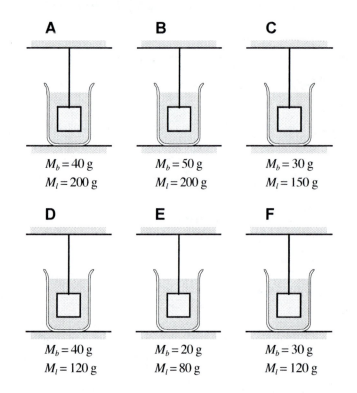

A	B	C
$M_b = 40$ g	$M_b = 50$ g	$M_b = 30$ g
$M_l = 200$ g	$M_l = 200$ g	$M_l = 150$ g

D	E	F
$M_b = 40$ g	$M_b = 20$ g	$M_b = 30$ g
$M_l = 120$ g	$M_l = 80$ g	$M_l = 120$ g

Greatest Force 1 _____ 2 _____ 3 _____ 4 _____ 5 _____ 6 _____ Smallest Force

Or, all of the blocks experience the same buoyant force. _____

Or, there is no buoyant force on any of these blocks. _____

Please carefully explain your reasoning.

How sure were you of your ranking? (circle one)

Basically Guessed				Sure				Very Sure	
1	2	3	4	5	6	7	8	9	10

[98] D. Maloney

Ranking Task Exercises in Physics: Student Edition   98

Blocks Suspended in Liquids—Volume of Liquid Displaced[99]

Shown below are six containers holding various liquids. Blocks of various solids are suspended in the liquids by being hung from a supporting rod. All of these blocks are the same size, but they have different masses (labeled as M_b) since they are made of different materials. All of the containers have the same volume of liquid, but the masses of these liquids vary (labeled M_l) since the liquids are different. Specific values for the masses of the blocks and the liquids are given in each figure. The volume of the blocks is one-fifth the volume of the liquids.

Rank these situations, from greatest to least, on the basis of the volume of the liquid displaced by the blocks.

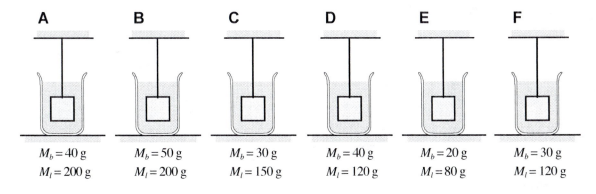

A	B	C	D	E	F
$M_b = 40$ g	$M_b = 50$ g	$M_b = 30$ g	$M_b = 40$ g	$M_b = 20$ g	$M_b = 30$ g
$M_l = 200$ g	$M_l = 200$ g	$M_l = 150$ g	$M_l = 120$ g	$M_l = 80$ g	$M_l = 120$ g

Greatest Volume 1 _____ 2 _____ 3 _____ 4 _____ 5 _____ 6 _____ Least Volume

Or, all of the volumes of the liquids displaced by the blocks are the same. _____

Please carefully explain your reasoning.

How sure were you of your ranking? (circle one)
Basically Guessed   Sure   Very Sure
1   2   3   4   5   6   7   8   9   10

Blocks Suspended in Liquids—Mass of Liquid Displaced[100]

Shown below are six containers holding various liquids. Blocks of various solids are suspended in the liquids by being hung from a supporting rod. All of these blocks are the same size, but they have different masses (labeled as M_b) since they are made of different materials. All of the containers have the same volume of liquid, but the masses of these liquids vary (labeled M_l) since the liquids are different. Specific values for the masses of the blocks and the liquids are given in each figure. The volume of the blocks is one-fifth the volume of the liquids.

Rank these situations, from greatest to least, on the basis of the mass of the liquids displaced by the blocks.

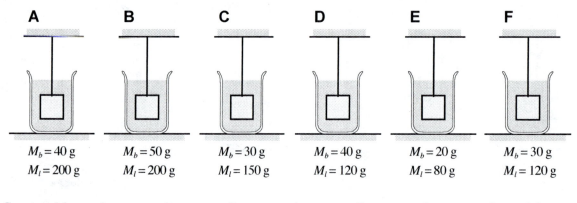

A	B	C	D	E	F
$M_b = 40$ g	$M_b = 50$ g	$M_b = 30$ g	$M_b = 40$ g	$M_b = 20$ g	$M_b = 30$ g
$M_l = 200$ g	$M_l = 200$ g	$M_l = 150$ g	$M_l = 120$ g	$M_l = 80$ g	$M_l = 120$ g

Greatest Mass   1 _____ 2 _____ 3 _____ 4 _____ 5 _____ 6 _____   Least Mass

Or, all of the masses of the liquids displaced by the blocks are the same. _____

Please carefully explain your reasoning.

How sure were you of your ranking? (circle one)

Basically Guessed   Sure   Very Sure

1   2   3   4   5   6   7   8   9   10

[100] D. Maloney, C. Hieggelke

Blocks Suspended in Liquids at Different Depths—Buoyant Force[101]

Shown below are six objects that have different masses and different volumes. These blocks are suspended at two different depths in water by being hung by a string from a supporting rod.

Rank these situations, from greatest to least, on the basis of buoyant force on the blocks by the water.

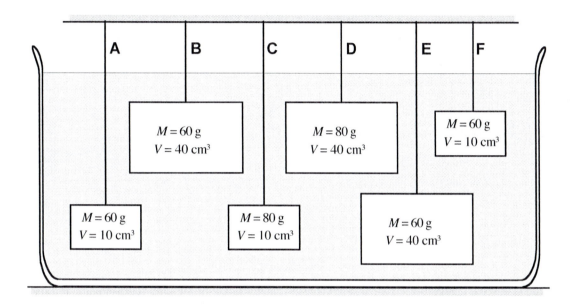

Greatest Force 1 _____ 2 _____ 3 _____ 4 _____ 5 _____ 6 _____ Least Force

Or, all of the buoyant forces on the blocks by the water are equal. _____

Or, there are no buoyant forces on the blocks by the water. _____

Or, it is not possible to determine the buoyant forces on the blocks by the water. _____

Please carefully explain your reasoning.

How sure were you of your ranking? (circle one)

Basically Guessed				Sure				Very Sure	
1	2	3	4	5	6	7	8	9	10

[101] C. Hieggelke

Ranking Task Exercises in Physics: Student Edition   101

Floating Blocks with Different Loads—Buoyant Force[102]

Shown below are six wood blocks which have different masses and different volumes. These blocks are floating in water. On top of these blocks are additional masses which provide a load for each of these blocks. Note: The blocks are not drawn correctly in terms of the depth the wooden blocks are in the water.

Rank these situations, from greatest to least, on the basis of buoyant force on the wood blocks by the water.

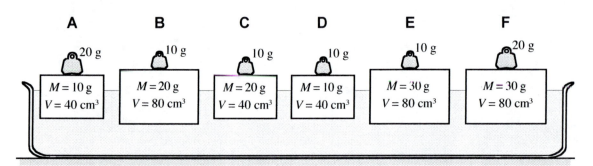

A   B   C   D   E   F

A: 20 g, $M = 10$ g, $V = 40$ cm³
B: 10 g, $M = 20$ g, $V = 80$ cm³
C: 10 g, $M = 20$ g, $V = 40$ cm³
D: 10 g, $M = 10$ g, $V = 40$ cm³
E: 10 g, $M = 30$ g, $V = 80$ cm³
F: 20 g, $M = 30$ g, $V = 80$ cm³

Greatest Force   1 _____ 2 _____ 3 _____ 4 _____ 5 _____ 6 _____   Least Force

Or, all of the buoyant forces on the wood blocks by the water are equal. _____

Or, there are no buoyant forces on the wood blocks by the water. _____

Or, it is not possible to determine the buoyant forces on the wood blocks by the water. ___

Please carefully explain your reasoning.

How sure were you of your ranking? (circle one)

Basically Guessed					Sure				Very Sure
1	2	3	4	5	6	7	8	9	10

Blocks Suspended in Water—String Tension[103]

Shown below are six containers that have the same volume of water in them. Blocks of various solids are suspended in the containers by being hung from a supporting rod. The blocks vary in both size and mass. The blocks are made of different materials, but all of the blocks would sink if the strings were cut. Specific values for the masses labeled as M_b and volumes labeled as V_b of the blocks are given in each figure.

Rank these situations, from greatest to least, on the basis of the tensions in the strings. That is, put first the situation that has the greatest tension in the string supporting the block, and put last the situation that has the lowest tension in the supporting string.

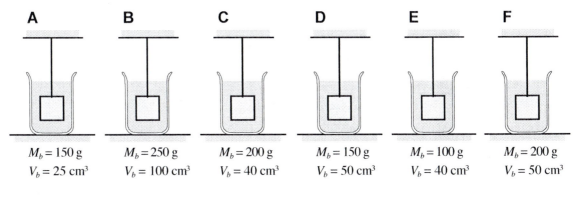

A	B	C	D	E	F
$M_b = 150$ g	$M_b = 250$ g	$M_b = 200$ g	$M_b = 150$ g	$M_b = 100$ g	$M_b = 200$ g
$V_b = 25$ cm³	$V_b = 100$ cm³	$V_b = 40$ cm³	$V_b = 50$ cm³	$V_b = 40$ cm³	$V_b = 50$ cm³

Greatest Tension 1 _____ 2 _____ 3 _____ 4 _____ 5 _____ 6 _____ Least Tension

Or, all of the strings have the same tension. _____

Or, the tension is zero in all of these strings. _____

Please carefully explain your reasoning.

How sure were you of your ranking? (circle one)
Basically Guessed   Sure   Very Sure
1   2   3   4   5   6   7   8   9   10

Blocks Suspended in Liquids—Buoyant Force II[104]

Shown below are six containers that have the same volume of the same liquid in them. Blocks of various solids are suspended in the containers by being hung from a supporting rod. The blocks vary in both size and mass. The blocks are made of different materials, but all of the blocks would sink if the strings were cut. Specific values for the masses labeled as M_b and volumes labeled as V_b of the blocks are given in each figure.

Rank these situations, from greatest to least, on the basis of the buoyant force on the blocks. That is, put first the situation that has the greatest buoyant force on the block, and put last the situation that has the lowest buoyant force on the block.

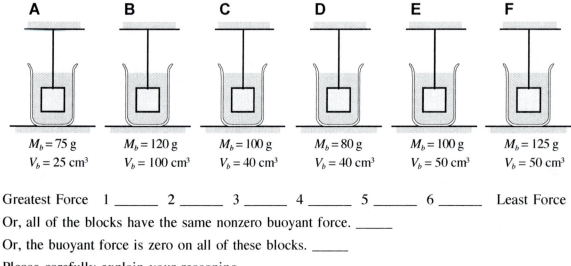

A	B	C	D	E	F
$M_b = 75$ g	$M_b = 120$ g	$M_b = 100$ g	$M_b = 80$ g	$M_b = 100$ g	$M_b = 125$ g
$V_b = 25$ cm³	$V_b = 100$ cm³	$V_b = 40$ cm³	$V_b = 40$ cm³	$V_b = 50$ cm³	$V_b = 50$ cm³

Greatest Force   1 _____ 2 _____ 3 _____ 4 _____ 5 _____ 6 _____ Least Force

Or, all of the blocks have the same nonzero buoyant force. _____

Or, the buoyant force is zero on all of these blocks. _____

Please carefully explain your reasoning.

How sure were you of your ranking? (circle one)

Basically Guessed					Sure				Very Sure	
1	2	3	4	5	6	7	8	9	10

[104] D. Maloney, C. Hieggelke

Blocks Floating in Liquids—Buoyant Force[105]

Shown below are six containers that have the same volume of the same liquid in them. Blocks of various solids are floating on top of the liquid. The blocks vary in both size and mass. Specific values for the masses labeled as M_b and volumes labeled as V_b of the blocks are given in each figure.

Rank these situations, from greatest to least, on the basis of the buoyant force by the liquid on the blocks. That is, put first the situation that has the greatest buoyant force by the liquid on the block, and put last the situation that has the lowest buoyant force by the liquid on the block.

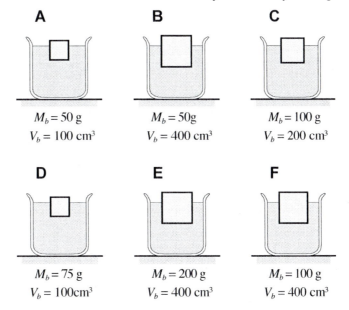

A	B	C
$M_b = 50$ g	$M_b = 50$g	$M_b = 100$ g
$V_b = 100$ cm³	$V_b = 400$ cm³	$V_b = 200$ cm³

D	E	F
$M_b = 75$ g	$M_b = 200$ g	$M_b = 100$ g
$V_b = 100$cm³	$V_b = 400$ cm³	$V_b = 400$ cm³

Greatest Force 1 _____ 2 _____ 3 _____ 4 _____ 5 _____ 6 _____ Least Force

Or, all of the blocks have the same buoyant force by the liquid. _____

Please carefully explain your reasoning.

How sure were you of your ranking? (circle one)

Basically Guessed				Sure				Very Sure	
1	2	3	4	5	6	7	8	9	10

[105] D. Maloney, C. Hieggelke
Ranking Task Exercises in Physics: Student Edition   105

Blocks at the Bottom of Liquids—Buoyant Force[106]

Shown below are six containers that have the same volume of the same liquid in them. Blocks of various solids are at the bottom of the containers. The blocks vary in both size and mass. Specific values for the masses labeled as M_b and volumes labeled as V_b of the blocks are given in each figure.

Rank these situations, from greatest to least, on the basis of buoyant force on the blocks. That is, put first the situation that has the greatest buoyant force on the block, and put last the situation that has the lowest buoyant force on the block.

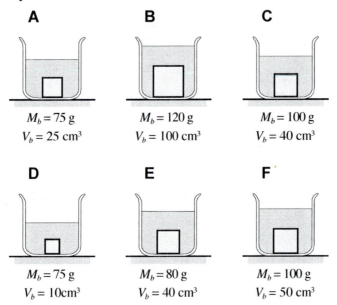

A

$M_b = 75$ g
$V_b = 25$ cm³

B

$M_b = 120$ g
$V_b = 100$ cm³

C

$M_b = 100$ g
$V_b = 40$ cm³

D

$M_b = 75$ g
$V_b = 10$ cm³

E

$M_b = 80$ g
$V_b = 40$ cm³

F

$M_b = 100$ g
$V_b = 50$ cm³

Greatest Force 1 _____ 2 _____ 3 _____ 4 _____ 5 _____ 6 _____ Least Force

Or, all of the blocks have the same buoyant force. _____

Or, the buoyant force is zero on all these blocks. _____

Please carefully explain your reasoning.

How sure were you of your ranking? (circle one)
Basically Guessed   Sure   Very Sure
1   2   3   4   5   6   7   8   9   10

Cylinders of Water—Pressure on the Plug I[107]

The figures below show side views of six hollow cylinders that have varying amounts of water in them. The widths of the cylinders and the heights to which they have been filled with water vary. The cylinders all have a hole cut in the side. All of the holes are the same size and they are all at the same height above the bases of the cylinders. There are corks in all of the holes.

Rank these situations, from greatest to smallest, on the basis of the pressure on the cork by the water.

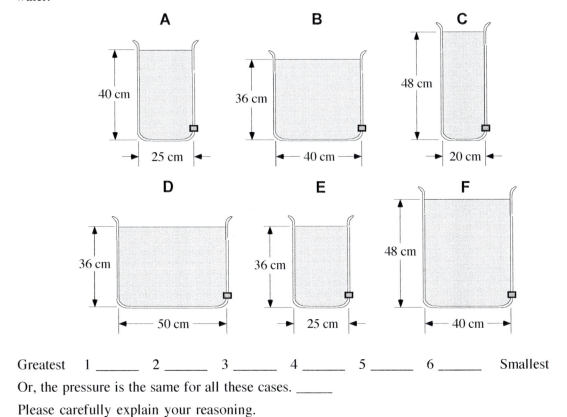

Greatest   1 _____   2 _____   3 _____   4 _____   5 _____   6 _____   Smallest

Or, the pressure is the same for all these cases. _____

Please carefully explain your reasoning.

How sure were you of your ranking? (circle one)

Basically guessed				Sure				Very Sure	
1	2	3	4	5	6	7	8	9	10

[107] D. Maloney, C. Hieggelke
Ranking Task Exercises in Physics: Student Edition   107

Cylinders of Liquids—Pressure on the Plug II[108]

The figures below show side views of six hollow cylinders that have varying amounts of various liquids in them. The widths of the cylinders and the heights to which they have been filled with liquid vary. The densities of the liquids, the widths of the cylinders, and the heights to which the cylinders have been filled are given in each figure. The cylinders all have a hole cut in the side. All of the holes are the same size and they are all at the same height above the bases of the cylinders. There are corks in all of the holes.

Rank these situations from greatest to smallest on the basis of the pressure on the cork by the water.

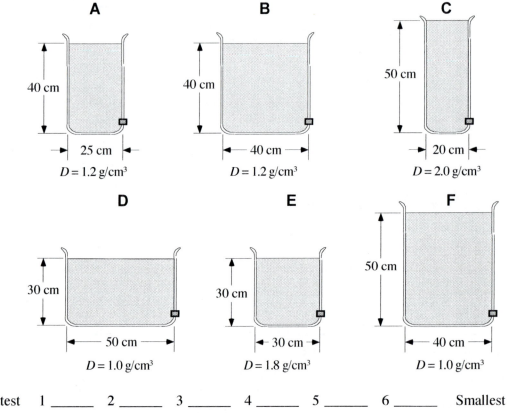

Greatest   1 _____   2 _____   3 _____   4 _____   5 _____   6 _____   Smallest

Or, the pressure on the cork by the water is the same for each case. _____

Please carefully explain your reasoning.

How sure were you of your ranking? (circle one)
Basically guessed   Sure   Very Sure
1   2   3   4   5   6   7   8   9   10

[108] D. Maloney, C. Hieggelke

Various Thermodynamic Processes—Work Done by Gas[109]

Six thermodynamic processes are illustrated below. All six processes are for the same ideal gas starting in the same state (same pressure, volume, temperature, and amount of gas). Each of the six distinctly different processes results in different final equilibrium states.

Rank these processes from greatest to least on the basis of the amount of work that is done by the gas. Positive work should be ranked higher than negative work.

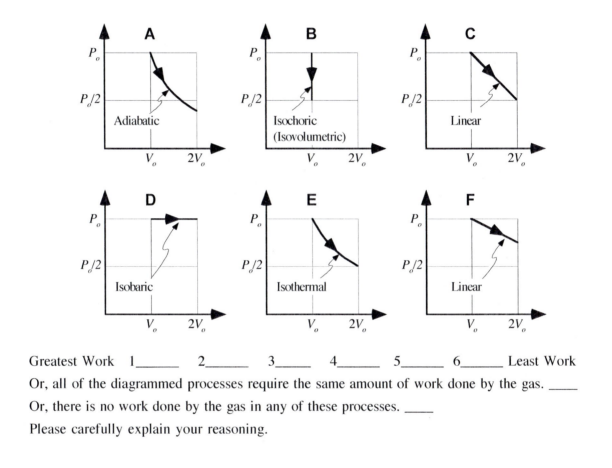

Greatest Work   1_____   2_____   3_____   4_____   5_____   6_____ Least Work

Or, all of the diagrammed processes require the same amount of work done by the gas. ____

Or, there is no work done by the gas in any of these processes. ____

Please carefully explain your reasoning.

[109] J. Popp

Gas Cylinders—Pressure[110]

Gas cylinders with the same mass (or number of molecules) of He gas have various volumes and temperatures, which are given in the diagram below. All of the containers are made of the same material.

Rank these containers from the highest to lowest on the basis of their gauge pressures.

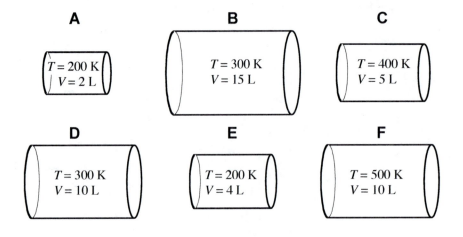

Highest   1 _____   2 _____   3 _____   4 _____   5 _____   6 _____   Lowest

Or, all of the containers have the same gauge pressure. _____

Please carefully explain your reasoning.

How sure were you of your ranking? (circle one)

Basically Guessed				Sure				Very Sure	
1	2	3	4	5	6	7	8	9	10

[110] U. Pandy, J. Splett, C. Hieggelke, T. O'Kuma, D. Maloney
Ranking Task Exercises in Physics: Student Edition   110

Pressure vs. Volume Graph—Temperature at Different Points I[111]

Rank from greatest to least the temperature of one mole of an ideal gas at the different points identified on the *P-V* diagram below.

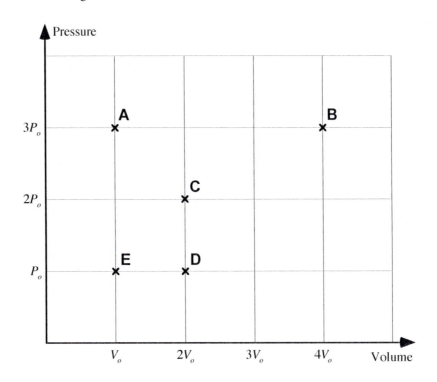

Greatest 1 _____ 2 _____ 3 _____ 4 _____ 5 _____ Least

Or, the gas has the same temperature at all these points._____

Or, it is not possible to rank the temperatures for these points._____

Please carefully explain your reasoning.

How sure were you of your ranking? (circle one)

Basically Guessed				Sure				Very Sure	
1	2	3	4	5	6	7	8	9	10

[111] J. Popp, C. Hieggelke

Pressure vs. Volume Graph—Temperature at Different Points II[112]

Rank from greatest to least the temperature of one mole of an ideal gas at the different points identified on the P-V diagram below.

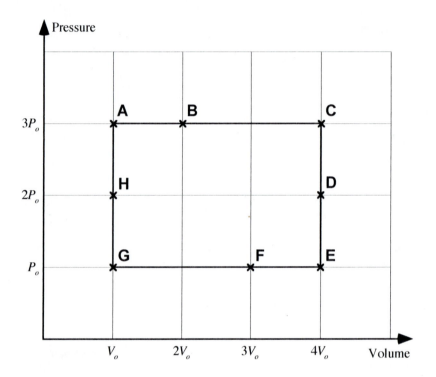

Greatest 1 _____ 2 _____ 3 _____ 4 _____ 5 _____ 6 _____ 7 _____ 8 _____ Least

Or, the gas has the same temperature at all these points._____

Or, it is not possible to rank the temperatures for these points._____

Please carefully explain your reasoning.

How sure were you of your ranking? (circle one)
Basically Guessed   Sure   Very Sure
1   2   3   4   5   6   7   8   9   10

[112] J. Popp, C. Hieggelke
Ranking Task Exercises in Physics: Student Edition   112

Pressure, Temperature, and Molecules—Internal Energy[113]

Rank from greatest to least the total internal energies of the ideal gases below that contain a different number (N) of molecules at various temperatures (T) and pressures (P). Note: The volumes are not given; they may or may not be the same.

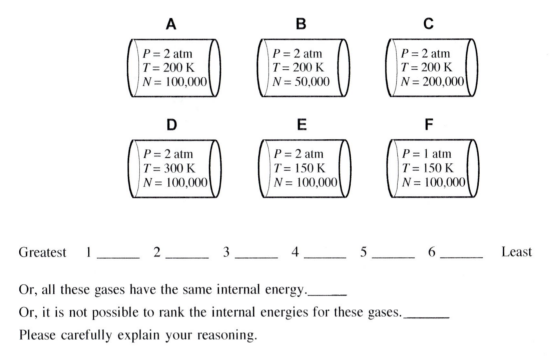

A

$P = 2$ atm
$T = 200$ K
$N = 100,000$

B

$P = 2$ atm
$T = 200$ K
$N = 50,000$

C

$P = 2$ atm
$T = 200$ K
$N = 200,000$

D

$P = 2$ atm
$T = 300$ K
$N = 100,000$

E

$P = 2$ atm
$T = 150$ K
$N = 100,000$

F

$P = 1$ atm
$T = 150$ K
$N = 100,000$

Greatest   1 _____   2 _____   3 _____   4 _____   5 _____   6 _____   Least

Or, all these gases have the same internal energy._____

Or, it is not possible to rank the internal energies for these gases._____

Please carefully explain your reasoning.

How sure were you of your ranking? (circle one)

Basically Guessed				Sure					Very Sure
1	2	3	4	5	6	7	8	9	10

Pressure, Volume, and Molecules—Temperature[114]

Rank from greatest to least the temperatures of the ideal gases below that contain a different number of molecules (N) at various pressures (P) and volumes (V).

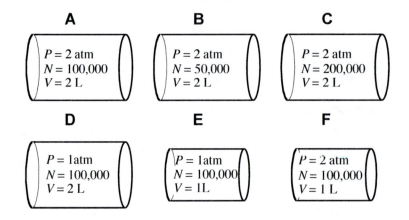

Greatest 1 _____ 2 _____ 3 _____ 4 _____ 5 _____ 6 _____ Least

Or, all these gases have the same temperature. _____

Or, it is not possible to rank the temperatures for these gases. _____

Please carefully explain your reasoning.

How sure were you of your ranking? (circle one)

Basically Guessed				Sure					Very Sure
1	2	3	4	5	6	7	8	9	10

[114] C. Hieggelke

Internal Energy, Volume, and Molecules—Temperature[115]

Rank from greatest to least the temperatures of the ideal gases below that contain various amounts of internal energy (U) and various numbers of molecules (N) in various volumes.

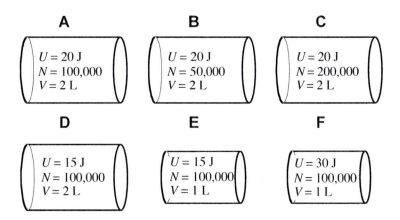

Greatest   1 _____ 2 _____ 3 _____ 4 _____ 5 _____ 6 _____   Least

Or, all these gases have the same temperature._____

Or, it is not possible to rank the temperatures for these gases._____

Please carefully explain your reasoning.

How sure were you of your ranking? (circle one)

Basically Guessed				Sure					Very Sure
1	2	3	4	5	6	7	8	9	10

[115] C. Hieggelke

Internal Energy, Volume, and Molecules—Pressure[116]

Rank from greatest to least the pressures of the ideal gases below that contain various amounts of internal energy (U) and various numbers of molecules (N) in various volumes (V).

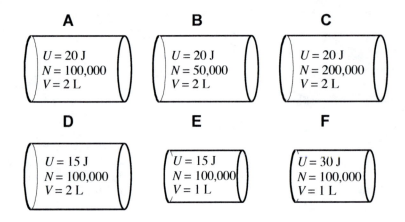

Greatest 1 _____ 2 _____ 3 _____ 4 _____ 5 _____ 6 _____ Least

Or, all these gases have the same pressure._____

Or, it is not possible to rank the pressures for these gases. _____

Please carefully explain your reasoning.

How sure were you of your ranking? (circle one)

Basically Guessed   Sure   Very Sure

1   2   3   4   5   6   7   8   9   10

[116] C. Hieggelke

Ranking Task Exercises in Physics: Student Edition   116

Temperature, Pressure, and Molecules—Internal Energy[117]

Rank from greatest to least the total internal energies of the ideal gases below that contain different numbers of molecules (N) at various temperatures (T) and pressures (P). Note: The volumes are not given.

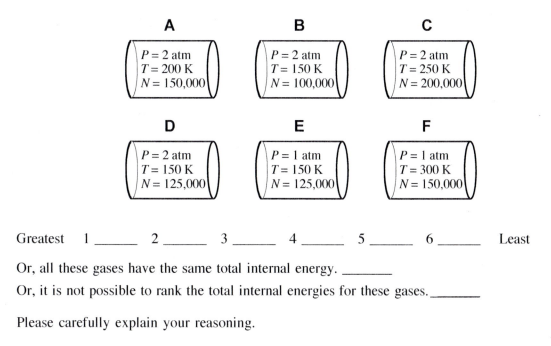

Greatest   1 _____   2 _____   3 _____   4 _____   5 _____   6 _____   Least

Or, all these gases have the same total internal energy. _____

Or, it is not possible to rank the total internal energies for these gases. _____

Please carefully explain your reasoning.

How sure were you of your ranking? (circle one)

Basically Guessed				Sure					Very Sure
1	2	3	4	5	6	7	8	9	10

[117] C. Hieggelke

Piston Containers—Pressure[118]

The figures below show piston containers with equal cross-sectional areas that contain different volumes of gas. The gases are the same and are at the same temperature. The pistons, which are free to move, are supporting various weights.

Rank these arrangements, from greatest to least, on the basis of the pressure of the gas in each container.

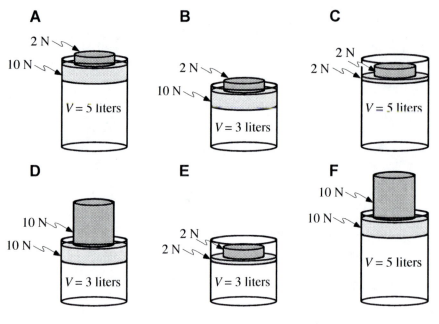

Greatest   1 _____   2 _____   3 _____   4 _____   5 _____   6 _____   Least

Or, all the containers have the same pressure. _____

Or, it is not possible to compare these pressures without knowing more. _____

Please carefully explain your reasoning.

How sure were you of your ranking? (circle one)

Basically Guessed					Sure				Very Sure
1	2	3	4	5	6	7	8	9	10

[118] C. Hieggelke, D. Maloney
Ranking Task Exercises in Physics: Student Edition   118

Piston Containers—Mass[119]

The figures below show piston containers with equal cross-sectional areas that contain different volumes of gas. The gases are the same and are at the same temperature. The pistons, which are free to move, are supporting various weights.

Rank these arrangements, from greatest to least, on the basis of the amount (mass) of gas in each container.

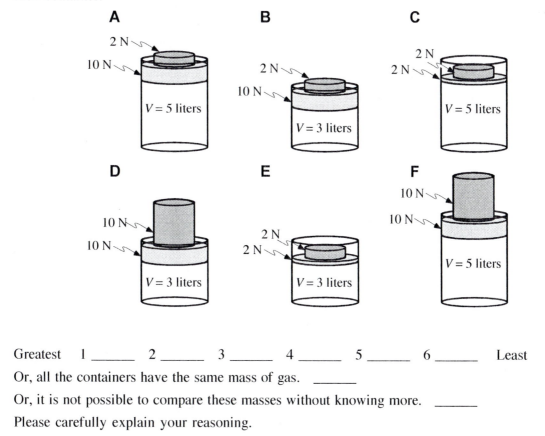

Greatest   1 _____   2 _____   3 _____   4 _____   5 _____   6 _____   Least

Or, all the containers have the same mass of gas. _____

Or, it is not possible to compare these masses without knowing more. _____

Please carefully explain your reasoning.

How sure were you of your ranking? (circle one)
Basically Guessed   Sure   Very Sure
1   2   3   4   5   6   7   8   9   10

[119] C. Hieggelke, D. Maloney
Ranking Task Exercises in Physics: Student Edition   119

Variety of Piston Containers—Pressure[120]

The figures below show piston containers with equal cross-sectional areas that contain different volumes of gas. The types of gases and their temperatures are indicated. The pistons, which are free to move, are supporting various weights.

Rank these arrangements, from greatest to least, on the basis of the pressure of the gas in each container.

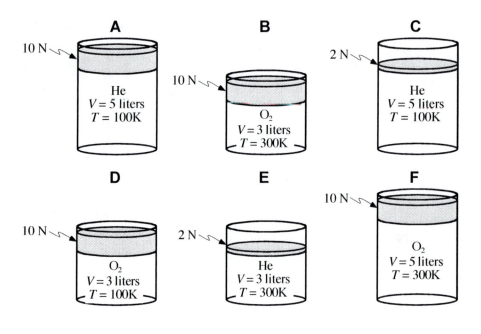

Greatest   1 _____   2 _____   3 _____   4 _____   5 _____   6 _____   Least

Or, all the containers have the same pressure. _____

Or, it is not possible to compare these pressures without knowing more. _____

Please carefully explain your reasoning.

How sure were you of your ranking? (circle one)

Basically Guessed					Sure				Very Sure	
1	2	3	4	5	6	7	8	9	10

[120] C. Hieggelke, D. Maloney

Variety of Piston Containers—Number of Moles[121]

The figures below show piston containers with equal cross-sectional areas that contain different volumes of gas. The types of gases and their temperatures are indicated. The pistons, which are free to move, are supporting various weights.

Rank these arrangements, from greatest to least, on the basis of the number of moles of gas in each container.

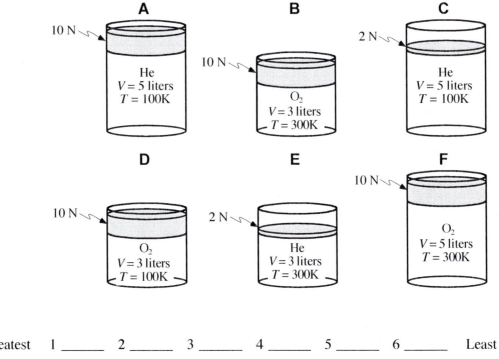

Greatest   1 _____   2 _____   3 _____   4 _____   5 _____   6 _____   Least

Or, all the containers have the same number of moles. _____

Or, it is not possible to compare these containers without knowing more. _____

Please carefully explain your reasoning.

How sure were you of your ranking? (circle one)

Basically Guessed   Sure   Very Sure

1   2   3   4   5   6   7   8   9   10

[121] C. Hieggelke, D. Maloney

Ranking Task Exercises in Physics: Student Edition   121

Two Electric Charges—Electric Force[122]

Given below are seven arrangements of two electric charges. In each figure, a point labeled P is also identified. All of the charges are the same size, $+Q$, but they can be either positive or negative. The charges and point P all lie on a straight line. The distances between adjacent items, either between two charges or between a charge and point P, are all x cm. There are no other charges in this region. For this problem, we are going to place a charge of equal magnitude at point P.

Rank these arrangements from strongest to weakest on the basis of the strength (magnitude) of the electric force on the charge when it is placed at point P. That is, put first the arrangement that will exert the strongest force on the charge at point P, and put last the arrangement that will exert the weakest force on the charge when it is placed at point P.

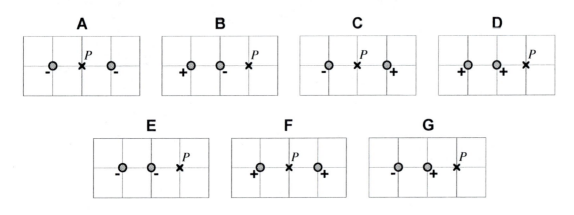

Strongest   1 _____   2 _____   3 _____   4 _____   5 _____   6 _____   7 _____   Weakest

Or, all of these arrangements exert the same strength force on the charge at P. _____

Or, all of these arrangements will exert zero force on the charge at P. _____

Please carefully explain your reasoning.

How sure were you of your ranking? (circle one)

Basically Guessed   Sure   Very Sure

1   2   3   4   5   6   7   8   9   10

Two Electric Charges—Electric Potential[123]

Given below are seven arrangements of two electric charges. In each figure, a point labeled P is also identified. All of the charges are the same size, q, but they can be either positive or negative. The charges and point P all lie on a straight line. The distances between adjacent items, either between two charges or between a charge and point P, are all d. There are no other charges in this region. For this problem there is no charge at point P.

Rank these arrangements, from greatest to least, on the basis of the electric potential at point P. That is, put first the arrangement that will have the greatest electric potential, and put last the arrangement that will have the least electric potential. Note that $-3 < 0 < 2$.

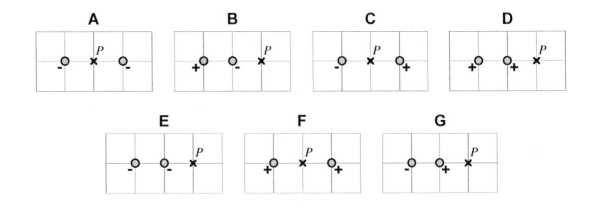

Greatest 1 _____ 2 _____ 3 _____ 4 _____ 5 _____ 6 _____ 7 _____ Least

Or, all of these arrangements exert the same electric potential. _____

Or, all of these arrangements will exert zero electric potential. _____

Please explain the reasoning for your ranking.

How sure were you of you ranking?
Basically Guessed   Sure   Very Sure
1   2   3   4   5   6   7   8   9   10

---

[123] T. O'Kuma

Three Electric Charges—Electric Potential Energy[124]

Given below are seven arrangements of two electric charges. In each figure, a point labeled P is also identified. All of the charges are the same size, q, but they can be either positive or negative. The charges and point P all lie on a straight line. The distances between adjacent items, either between two charges or between a charge and point P, are all d. There are no other charges in this region. For this problem we are going to place a $+2q$ charge at point P.

Rank these arrangements, from greatest to least, on the basis of the electric potential energy of the system. That is, put first the arrangement that will have the greatest electric potential energy, and put last the arrangement that will have the least electric potential energy. Note that $-3 < 0 < 2$.

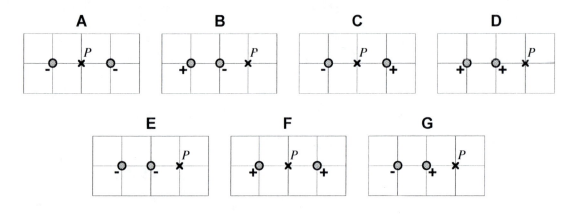

Greatest   1 _____   2 _____   3 _____   4 _____   5 _____   6 _____   7 _____   Least

Or, all of these arrangements exert the same electric potential energy for the system. _____

Or, all of these arrangements will exert zero electric potential energy for the system. _____

Please explain the reasoning for your ranking.

How sure were you of you ranking?
Basically Guessed   Sure   Very Sure
1   2   3   4   5   6   7   8   9   10

Three Linear Electric Charges—Electric Force[125]

Given below are arrangements of three fixed electric charges. In each figure, a point labeled P is also identified. All of the charges are the same size charge, q, but they can be either positive or negative as indicated. The charges and point P all lie on a straight line. The distances between adjacent items, either between two charges or between a charge and point P, are all the same. There are no other charges in this region. A test charge, $+Q$, is placed at point P.

Rank these arrangements from greatest to least on the basis of the strength (magnitude) of the electric force on the test charge, $+Q$, at P.

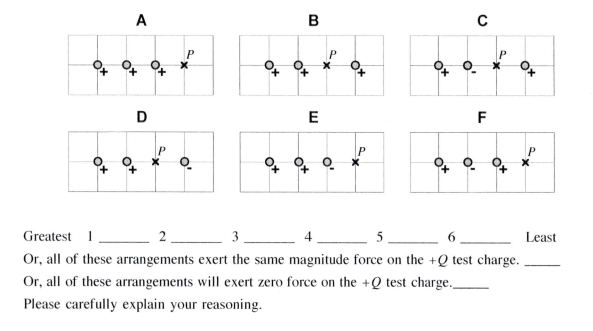

Greatest 1 _____ 2 _____ 3 _____ 4 _____ 5 _____ 6 _____ Least

Or, all of these arrangements exert the same magnitude force on the $+Q$ test charge. _____

Or, all of these arrangements will exert zero force on the $+Q$ test charge._____

Please carefully explain your reasoning.

How sure were you of your ranking? (circle one)

Basically Guessed				Sure				Very Sure	
1	2	3	4	5	6	7	8	9	10

[125] T. O'Kuma

Two Nonlinear Electric Charges—Electric Force[126]

Given below are arrangements of two fixed electric charges. In each figure, a point labeled P is also identified. All of the charges are the same size, q, but they can be either positive or negative as indicated. The distances between adjacent items, either between two charges or between a charge and point P, are all the same. There are no other charges in this region. For this problem, we are going to place a test charge, $+Q$, at point P.

Rank these arrangements from greatest to least on the basis of the strength (magnitude) of the electric force on the test charge, $+Q$, at P.

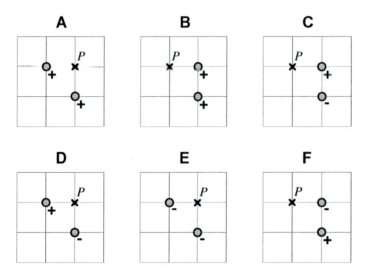

Greatest   1 _____ 2 _____ 3 _____ 4 _____ 5 _____ 6 _____ Least

Or, all of these arrangements exert the same magnitude force on the $+Q$ test charge._____

Or, all of these arrangements will exert zero force on the $+Q$ test charge._____

Please carefully explain your reasoning.

How sure were you of your ranking? (circle one)

Basically Guessed   Sure   Very Sure

  1   2   3   4   5   6   7   8   9   10

Charged Conducting Spheres—Electric Field at the Center[127]

Shown below are six hollow spheres of different sizes made of an electrically conducting material such as copper. On each sphere there is a charge, as given in the figure, which is distributed evenly over the sphere. Each figure is independent of the others (they do not affect each other).

Rank these situations, from greatest to least, on the basis of the magnitude of the electric field at the center of the sphere.

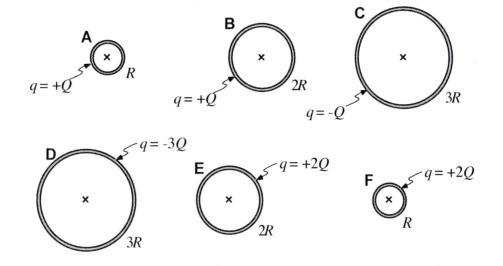

Greatest   1 _____   2 _____   3 _____   4 _____   5 _____   6 _____   Least

Or, the magnitude of the electric field at the center is the same (but not zero) for these cases. ___

Or, the magnitude of the electric field at the center is zero for these cases. _____

Please carefully explain your reasoning.

How sure were you of your ranking? (circle one)

Basically Guessed   Sure   Very Sure

1   2   3   4   5   6   7   8   9   10

_____

[127] C. Hieggelke, T. O'Kuma

Ranking Task Exercises in Physics: Student Edition   127

Charged Conducting Spheres—Electric Potential at the Center[128]

Shown below are six hollow spheres of different sizes made of an electrically conducting material such as copper. On each sphere there is a charge, as given in the figure, which is distributed evenly over the sphere. Each figure is independent of the others (they do not affect each other).

Rank these situations, from greatest to least, on the basis of the electric potential at the center of the sphere.

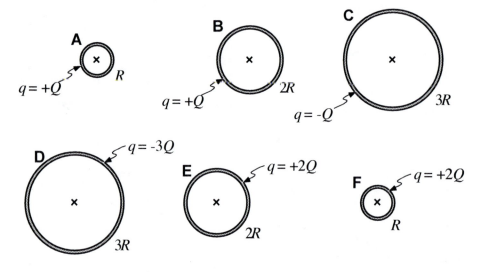

Greatest 1 _____ 2 _____ 3 _____ 4 _____ 5 _____ 6 _____ Least

Or, the electric potential at the center is the same (but not zero) for these cases. _____

Or, the electric potential at the center is zero for these cases. _____

Please carefully explain your reasoning.

How sure were you of your ranking? (circle one)

Basically Guessed				Sure				Very Sure	
1	2	3	4	5	6	7	8	9	10

[128] C. Hieggelke, T. O'Kuma
Ranking Task Exercises in Physics: Student Edition   128

Charged Conducting Spheres—Electric Field at Various Points[129]

Shown below are two hollow spheres made of an electrically conducting material such as copper. Each sphere has a different charge, which is given in the figure, distributed evenly over the sphere. Each figure or sphere is independent of the others (they do not affect each other).

Rank these situations, from greatest to least, on the basis of the magnitude of the electric field at the following points: **A** and **D,** which are inside the sphere at the center of the spheres; **B** and **E,** which are inside the sphere at a distance of R/2 from the center of the spheres; and **C** and **F,** which are outside the sphere at a distance of 3R/2 from the center of the spheres.

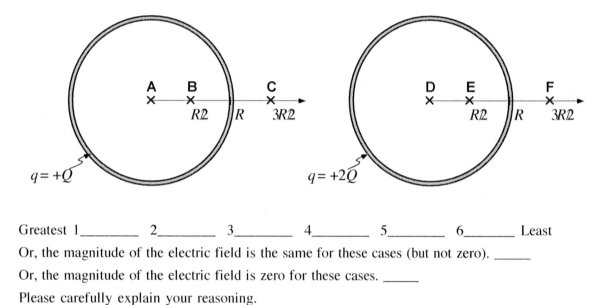

Greatest 1_____ 2_____ 3_____ 4_____ 5_____ 6_____ Least

Or, the magnitude of the electric field is the same for these cases (but not zero). _____

Or, the magnitude of the electric field is zero for these cases. _____

Please carefully explain your reasoning.

[129] C. Hieggelke, T. O'Kuma

Charged Conducting Spheres—Electric Potential at Various Points[130]

Shown below are two hollow spheres made of an electrically conducting material such as copper. Each sphere has a different charge, which is given in the figure, distributed evenly over the sphere. Each figure or sphere is independent of the others (they do not affect each other).

Rank these situations, from greatest to least, on the basis of the electric potential at the following points: **A** and **D**, which are inside the sphere at the center of the spheres; **B** and **E**, which are inside the sphere at a distance of R/2 from the center of the spheres; and **C** and **F**, which are outside the sphere at a distance of 3R/2 from the center of the spheres.

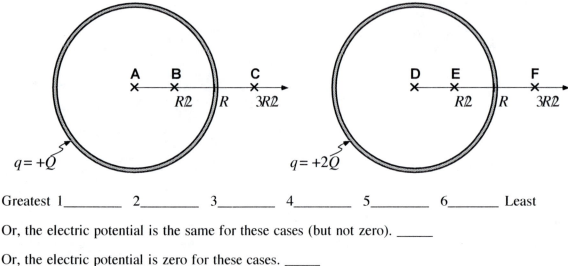

Greatest 1_____ 2_____ 3_____ 4_____ 5_____ 6_____ Least

Or, the electric potential is the same for these cases (but not zero). _____

Or, the electric potential is zero for these cases. _____

Please carefully explain your reasoning.

How sure were you of your ranking? (circle one)
Basically Guessed   Sure   Very Sure
1   2   3   4   5   6   7   8   9   10

[130] C. Hieggelke, T. O'Kuma
Ranking Task Exercises in Physics: Student Edition   130

Charges Outside a Conducting Sphere—Force on Enclosed Charge[131]

Shown below are six hollow spheres made of an electrically conducting material such as copper. At the center of each sphere is a charge q, which is the same sign and magnitude for all six cases. Outside these spheres, at various distances, are electric charges of various magnitudes. Given in each figure is the magnitude of the outside charges, as well as the distance between that charge and the sphere. Each figure is independent of the others (they do not affect each other).

Rank these situations, from greatest to least, on the basis of the magnitude of the force on the charge at the center of the sphere. That is, put first the situation where the charge at the center of the sphere experiences the strongest force, and put last the situation where the charge in the center experiences the weakest force.

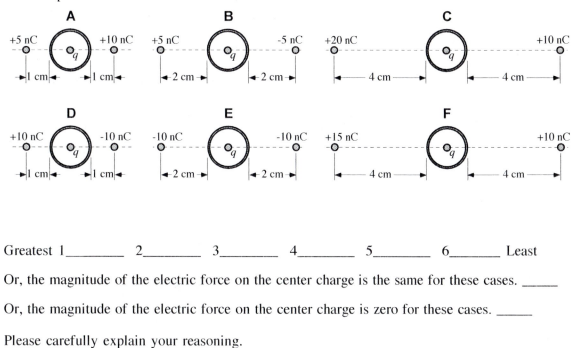

Greatest 1_____ 2_____ 3_____ 4_____ 5_____ 6_____ Least

Or, the magnitude of the electric force on the center charge is the same for these cases. _____

Or, the magnitude of the electric force on the center charge is zero for these cases. _____

Please carefully explain your reasoning.

How sure were you of your ranking? (circle one)
Basically Guessed   Sure   Very Sure
1   2   3   4   5   6   7   8   9   10

[131] C. Hieggelke, D. Maloney, T. O'Kuma
Ranking Task Exercises in Physics: Student Edition   131

Point Charges Outside Conducting Spheres—Electric Field Within[132]

Shown below are six identical hollow spheres with a radius of 1 cm made of an electrically conducting material such as copper. Outside these spheres, at various distances, are two electric charges of various magnitudes. Some of the charges are positive and some are negative. Given in each figure is the sign and magnitude of the charge, as well as the distance of the charge from the sphere. Each figure is independent of the others (they do not affect each other).

Rank these situations, from greatest to least, on the basis of the strengths (magnitudes) of the electric fields at the centers of the spheres.

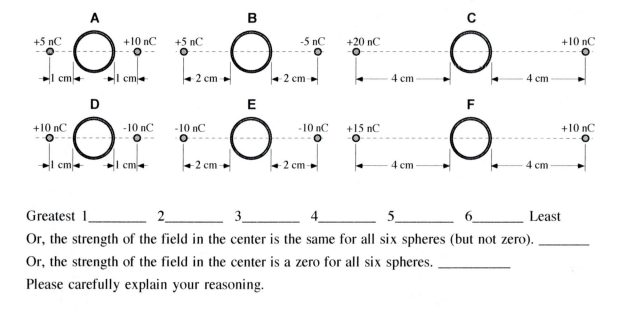

Greatest 1_____ 2_____ 3_____ 4_____ 5_____ 6_____ Least

Or, the strength of the field in the center is the same for all six spheres (but not zero). _____

Or, the strength of the field in the center is a zero for all six spheres. _____

Please carefully explain your reasoning.

How sure were you of your ranking? (circle one)

Basically Guessed				Sure				Very Sure	
1	2	3	4	5	6	7	8	9	10

Point Charges near Uncharged Conductors – Attractive Force[133]

In each of the six situations shown below, identical positive point charges are brought to within a distance d of identical oval-shaped conductors. All of the conductors were initially uncharged. The ovals labeled B and E are touching the adjacent oval, but there is a small gap between both C and F and the oval to their left.

Rank these arrangements, from strongly attracted to strongly repelled, on the basis of the strength (magnitude) of the electric *attraction* of the labeled oval to the point charge after the point charge is brought to the position shown. That is, put first the oval that is most strongly attracted by the point charge and put last the oval that is most strongly repelled.

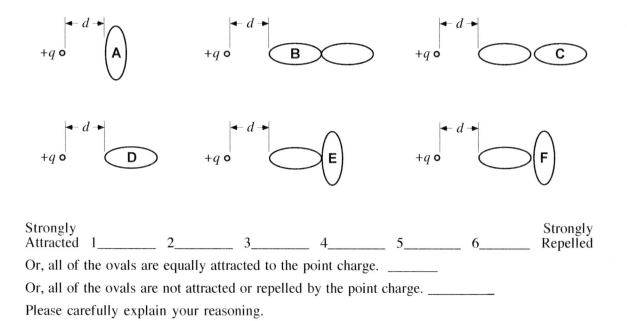

Strongly
Attracted   1_____   2_____   3_____   4_____   5_____   6_____   Strongly Repelled

Or, all of the ovals are equally attracted to the point charge. _____

Or, all of the ovals are not attracted or repelled by the point charge. _____

Please carefully explain your reasoning.

How sure were you of your ranking? (circle one)
Basically Guessed   Sure   Very Sure
1   2   3   4   5   6   7   8   9   10

[133] S. Kanim

Induced Charges—Near a Charged Rod[134]

A (positively or negatively) charged rod is brought up to the same distance from each set of metal spheres as shown in separate situations below. The spheres in each pair are initially in contact, but they are then separated while the rod is still in place. Then the rod is removed.

Rank the net charge on each sphere from most positive to most negative after the spheres have been separated and the charged rod removed.

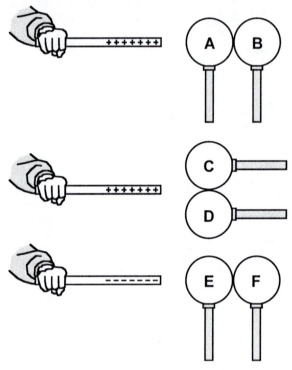

Most
Positive   1_____ 2_____ 3_____ 4_____ 5_____ 6_____   Most
Negative

Or, all spheres have the same charge. _____

Please carefully explain your reasoning.

How sure were you of your ranking? (circle one)
Basically Guessed   Sure   Very Sure
1   2   3   4   5   6   7   8   9   10

Pairs of Charges—Attractive and Repulsive Force[135]

The following diagrams show three separate pairs of point charges. The pairs only interact with each other since they are far apart.

Rank the force on each point charge from most attractive to most repulsive.

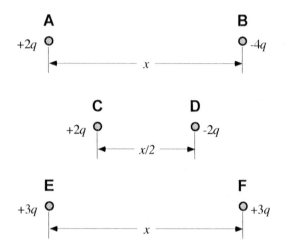

Most
Attracted 1_____ 2_____ 3_____ 4_____ 5_____ 6_____ Repelled

Most
                                                                              Repelled

Or, all of these charges experience the same force. _____

Please carefully explain your reasoning.

How sure were you of your ranking? (circle one)
Basically Guessed   Sure   Very Sure
1   2   3   4   5   6   7   8   9   10

[135] B. Emerson

Suspended Charges—Angle[136]

The figure below shows a charged sphere that is suspended from a string in a uniform electric field that is pointing in the horizontal direction. Six possible combinations of sphere mass and electric charge are listed in the chart below. All of these spheres are suspended in the same uniform field and the gravitational field is directed downward. When the spheres are placed in the field they will swing away from vertical and hang at an angle θ from the vertical as shown in the diagram.

Rank, from greatest to least, the angle θ the string will form with the vertical for these different spheres.

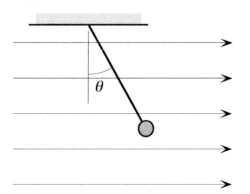

	Mass	Charge
A	3 g	8 nC
B	6 g	4 nC
C	9 g	2 nC
D	3 g	2 nC
E	6 g	8 nC
F	9 g	4 nC

Greatest   1 _____ 2 _____ 3 _____ 4 _____ 5 _____ 6 _____   Least

Or, all of the angles will be the same. _____

Please carefully explain your reasoning.

How sure were you of your ranking? (circle one)
Basically Guessed   Sure   Very Sure
1   2   3   4   5   6   7   8   9   10

[136] G. Fazzari, A. Weiner, T. Bennie-George, O. Selgrad, C. Hieggelke

Uniform Electric Field—Electric Force on Charge at Rest I[137]

We have a large region of space that has a uniform electric field in the +x direction (⇒). At the point (0,0) m, the electric field is 30 **i** N/C and the electric potential is 100 volts.

Rank from greatest to least the strength (magnitude) of the electric force on a +5 C charge when it is placed at rest at each of the following points.

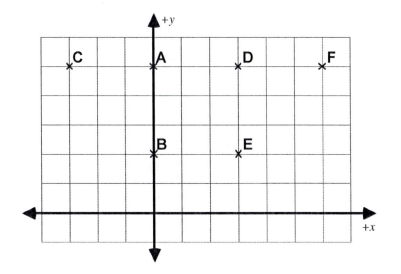

	x	y
A	0 m	5 m
B	0 m	2 m
C	-3 m	5 m
D	3 m	5 m
E	3 m	2 m
F	6 m	5 m

Greatest   1 _____   2 _____   3 _____   4 _____   5 _____   6 _____   Least

Or, the 5 C charge will experience the same strength electric force at all of these points. _____

Or, the 5 C charge will not experience a force at any of these points. _____

Please carefully explain your reasoning.

How sure were you of your ranking? (circle one)

Basically Guessed				Sure				Very Sure	
1	2	3	4	5	6	7	8	9	10

Uniform Electric Field—Electric Force on Charge at Rest II[138]

We have a large region of space that has a uniform electric field in the +x direction (⇒) as indicated by the arrows in the diagram below. At the point (0,0) m, the electric field is 30 **i** N/C and the electric potential is 100 volts.

Rank from greatest to least the strength (magnitude) of the electric force on a +5 C charge when it is placed at rest at each of the following points.

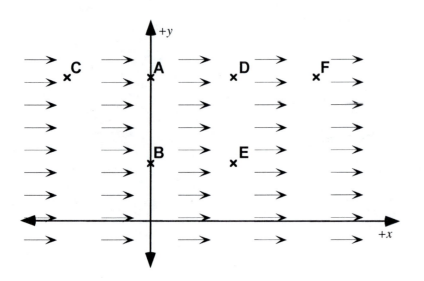

	x	y
A	0 m	5 m
B	0 m	2 m
C	-3 m	5 m
D	3 m	5 m
E	3 m	2 m
F	6 m	5 m

Greatest   1 _____   2 _____   3 _____   4 _____   5 _____   6 _____   Least

Or, the 5 C charge will experience the same strength electric force at all of these points. _____

Or, the 5 C charge will not experience a force at any of these points. _____

Please carefully explain your reasoning.

How sure were you of your ranking? (circle one)

Basically Guessed					Sure			Very Sure	
1	2	3	4	5	6	7	8	9	10

Uniform Electric Field/Potential Lines—Force on Charge at Rest III[139]

We have a large region of space that has a uniform electric field in the +x direction (⇒). In the diagram below we show the equipotential lines for this field. At the point (0,0) m, the electric field is 30 **i** N/C and the electric potential is 100 volts.

Rank from greatest to least the strength (magnitude) of the electric force on a +5 C charge when it is placed at rest at each of the following points.

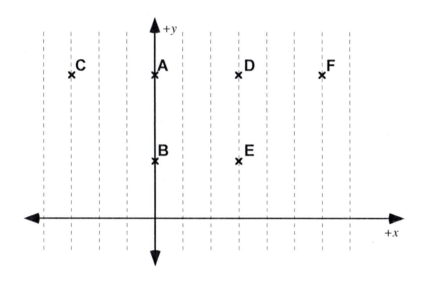

	x	y
A	0 m	5 m
B	0 m	2 m
C	-3 m	5 m
D	3 m	5 m
E	3 m	2 m
F	6 m	5 m

Greatest   1 _____   2 _____   3 _____   4 _____   5 _____   6 _____   Least

Or, the 5 C charge will experience the same strength electric force at all of these points. _____

Or, the 5 C charge will not experience a force at any of these points. _____

Please carefully explain your reasoning.

How sure were you of your ranking? (circle one)

Basically Guessed   Sure   Very Sure
1   2   3   4   5   6   7   8   9   10

Uniform Electric Field—Electric Force on Various Charges at Rest[140]

We have a large region of space that has a uniform electric field in the +x direction (⇒) as indicated by the arrows in the diagram below. At the point (0,0) m, the electric field is 30 **i** N/C and the electric potential is 100 volts.

Rank from greatest to least the strength (magnitude) of the electric force on the charges listed when placed at rest at the points specified below. Each charge is placed at its specified point separately.

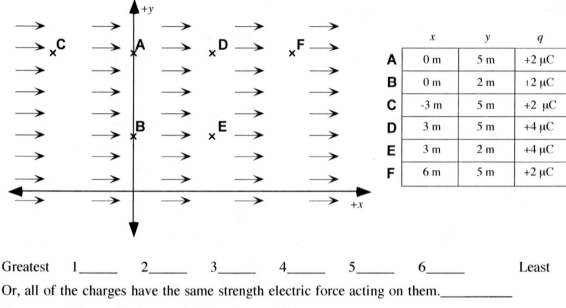

	x	y	q
A	0 m	5 m	+2 μC
B	0 m	2 m	ι2 μC
C	-3 m	5 m	+2 μC
D	3 m	5 m	+4 μC
E	3 m	2 m	+4 μC
F	6 m	5 m	+2 μC

Greatest   1_____   2_____   3_____   4_____   5_____   6_____   Least

Or, all of the charges have the same strength electric force acting on them._____

Or, none of these charges would experience a force. _____

Please carefully explain your reasoning.

How sure were you of your ranking? (circle one)
Basically Guessed   Sure   Very Sure
1   2   3   4   5   6   7   8   9   10

Two Electric Charges—Electric Field Along Line[141]

Given below are seven arrangements of two electric charges. In each figure, a point labeled *P* is also identified. All of the charges are the same size, *Q*, but they can be either positive or negative. The charges and point *P* all lie on a straight line. The distances between adjacent items, either between two charges or between a charge and point *P*, are all x cm. There is no charge at point *P*, nor are there any other charges in this region.

Rank these arrangements from strongest to weakest on the basis of the strength (magnitude) of the electric field at point *P*. That is, put first the arrangement that produces the strongest field at point *P*, and put last the arrangement that produces the weakest field at point *P*.

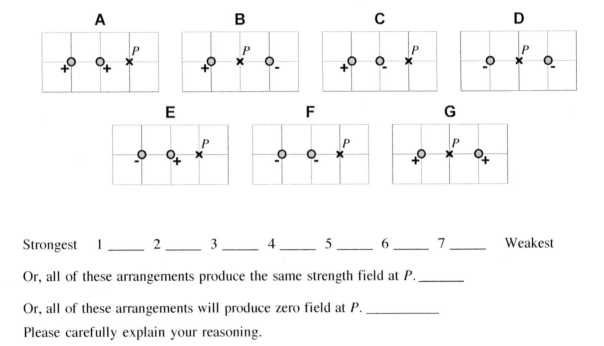

Strongest   1 _____   2 _____   3 _____   4 _____   5 _____   6 _____   7 _____   Weakest

Or, all of these arrangements produce the same strength field at *P*. _____

Or, all of these arrangements will produce zero field at *P*. _____

Please carefully explain your reasoning.

How sure were you of your ranking? (circle one)
Basically Guessed   Sure   Very Sure
1   2   3   4   5   6   7   8   9   10

[141] D. Maloney
Ranking Task Exercises in Physics: Student Edition   141

Electron Within a Charged Capacitor—Force on the Electron[142]

Consider an electron between the plates of a charged capacitor. The figures below show situations where the potential across the capacitor and the separations between the capacitor plates vary. Specific values are given in each figure.

Rank from greatest to least according to the magnitude of the force felt by the electron.

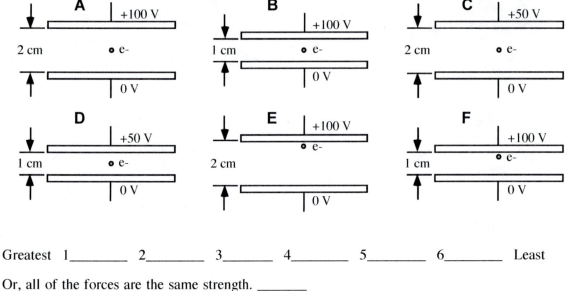

Greatest 1_____ 2_____ 3_____ 4_____ 5_____ 6_____ Least

Or, all of the forces are the same strength. _____

Please carefully explain your reasoning._____

How sure were you of your ranking? (circle one)

Basically Guessed				Sure				Very Sure	
1	2	3	4	5	6	7	8	9	10

[142] A. Van Heuvelen, S. Heath, B. Willis, L. Bryant

Transfer of Charge in Conductors—Left Conductor[143]

In each of the following situations two conducting spheres of the same size are shown with an initial given number of units of charge. The two spheres are brought into contact with each other. After several moments the spheres are separated.

Rank the situations as to the quantity of charge on the first (or left) sphere from the highest positive charge to the lowest negative charge after they have been separated. (Note that -6 is lower than –2.)

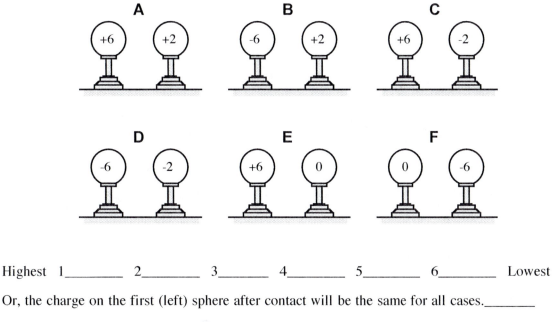

Highest 1_____ 2_____ 3_____ 4_____ 5_____ 6_____ Lowest

Or, the charge on the first (left) sphere after contact will be the same for all cases._____

Or, the first (left) sphere after contact will have no charge for all cases._____

Please carefully explain your reasoning.

How sure were you of your ranking? (circle one)
Basically Guessed   Sure   Very Sure
1   2   3   4   5   6   7   8   9   10

[143] J. Gundlach, B. Kaasa, U. Pandey, M. West

Transfer of Charge in Conductors—Right Conductor[144]

In each of the following situations two conducting spheres of the same size are shown with an initial given number of units of charge. The two spheres are brought into contact with each other. After several moments the spheres are separated.

Rank the situations as to the quantity of charge on the second (right) sphere from the highest positive charge to the lowest negative charge after they have been separated. (Note that -6 is lower than –2.)

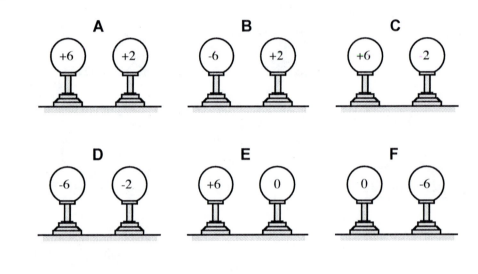

Highest 1_____ 2_____ 3_____ 4_____ 5_____ 6_____ Lowest

Or, the charge on the second (right) sphere after contact will be the same for all cases._____

Or, the second (right) sphere after contact will have no charge for all cases.____

Please carefully explain your reasoning.

How sure were you of your ranking? (circle one)
Basically Guessed   Sure   Very Sure
1   2   3   4   5   6   7   8   9   10

----
[144] J. Gundlach, B. Kaasa, U. Pandey, M. West
Ranking Task Exercises in Physics: Student Edition   144

Non-Uniform Electric Field—Electric Potential[145]

A non-uniform electric field is represented in the figure below by electric field lines. Six points in this region are identified in this diagram.

Rank the electric potential of the marked points from greatest to least.

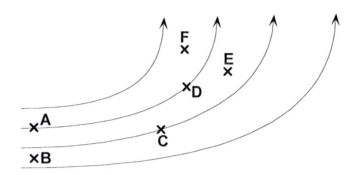

Greatest   1_____   2_____   3_____   4_____   5_____   6_____   Least

Or, all of these points have the same electric potentials._____

Please carefully explain your reasoning.

How sure were you of your ranking? (Circle one)
Basically Guessed   Sure   Very Sure
1   2   3   4   5   6   7   8   9   10

[145] R. Johnson, B. Keramati, C. Lam, S. Savrda
Ranking Task Exercises in Physics: Student Edition   145

Non-Uniform Electric Field—Electric Field Strength[146]

A non-uniform electric field is represented in the figure below by electric field lines. Six points in this region are identified in this diagram.

Rank the strength (magnitude) of the electric field of the marked points from greatest to least.

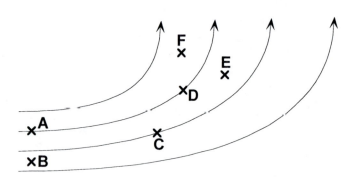

Greatest 1_____ 2_____ 3_____ 4_____ 5_____ 6_____ Least

Or, the magnitude of the electric field is the same at all of these points._____

Please carefully explain your reasoning.

How sure were you of your ranking? (circle one)
Basically Guessed   Sure   Very Sure
1   2   3   4   5   6   7   8   9   10

_____
[146] R. Johnson, B. Keramati, C. Lam, S. Savrda

Uniform Electric Field—Potential Difference[147]

Two parallel plates that have been charged create a uniform electric field of 30 N/C between the plates.

Rank from highest to lowest the electrical potential differences of all the different combinations listed below between the four points - **M** at (2,0) m; **N** at (5, 0)m; **O** at (8,0) m; and **P** at (2, 3) m within this region. (Positive values are larger than negative values.)

	From	To
A	M	N
B	M	O
C	N	O
D	P	M
E	P	N
F	P	O
G	N	M

Highest   1 _____ 2 _____ 3 _____ 4 _____ 5 _____ 6 _____ 7 _____   Lowest

Or, all the combinations have the same potential difference._____

Please carefully explain your reasoning.

How sure were you of your ranking? (circle one)
Basically Guessed   Sure   Very Sure
1   2   3   4   5   6   7   8   9   10

[147] A. Dickison, C. Ezrailson, M. Plumb, D. Ting

Uniform Electric Field—Strength of the Electric Field I[148]

We have a large region of space that has a uniform electric field in the +x direction (⇒). At the point (0,0) m, the electric field is 30 **i** N/C and the electric potential is 100 volts.

Rank the strength (magnitude) of the electric field from greatest to least at the following points within this region.

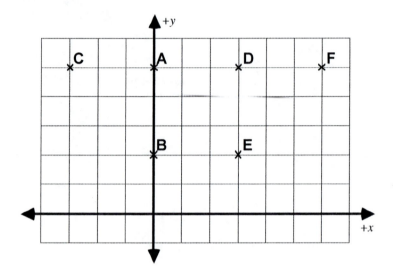

	x	y
A	0 m	5 m
B	0 m	2 m
C	-3 m	5 m
D	3 m	5 m
E	3 m	2 m
F	6 m	5 m

Greatest   1_____   2_____   3_____   4_____   5_____   6_____   Least

Or, all of the points have the same electric field magnitude._____

Please carefully explain your reasoning.

How sure were you of your ranking? (circle one)
Basically Guessed   Sure   Very Sure
1   2   3   4   5   6   7   8   9   10

Uniform Electric Field—Strength of the Electric Field II[149]

We have a large region of space that has a uniform electric field in the +*x* direction (⇒) as indicated by the arrows in the diagram below. At the point (0,0) m, the electric field is 30 **i** N/C and the electric potential is 100 volts.

Rank the strength (magnitude) of the electric field from greatest to least at the following points within this region.

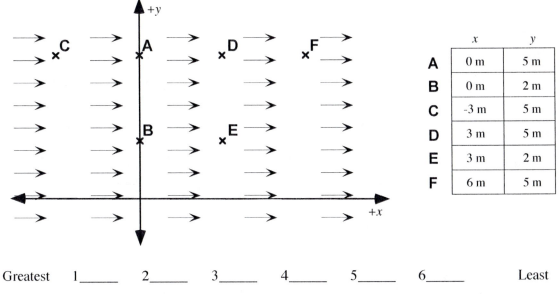

	x	*y*
A	0 m	5 m
B	0 m	2 m
C	-3 m	5 m
D	3 m	5 m
E	3 m	2 m
F	6 m	5 m

Greatest   1_____   2_____   3_____   4_____   5_____   6_____   Least

Or, all of the points have the same electric field magnitude._____

Please carefully explain your reasoning.

Uniform Electric Field/Potential Lines—Strength of the Electric Field[150]

We have a large region of space that has a uniform electric field in the +x direction ($\Rightarrow$). In the diagram below we show the equipotential lines for this field. At the point (0,0) m, the electric field is 30 **i** N/C and the electric potential is 100 volts.

Rank the strength (magnitude) of the electric field from greatest to least at the following points within this region.

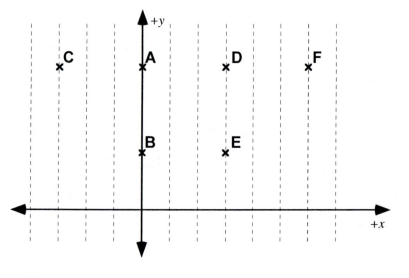

	x	y
A	0 m	5 m
B	0 m	2 m
C	-3 m	5 m
D	3 m	5 m
E	3 m	2 m
F	6 m	5 m

Greatest   1_____   2_____   3_____   4_____   5_____   6_____   Least

Or, all of the points have the same electric field magnitude._____

Please carefully explain your reasoning.

How sure were you of your ranking? (circle one)

Basically Guessed   Sure   Very Sure

1   2   3   4   5   6   7   8   9   10

Uniform Electric Field—Electric Potential at Different Points II[151]

We have a large region of space that has a uniform electric field in the $+x$ direction ($\Rightarrow$) as indicated by the arrows in the diagram below. At the point (0,0) m, the electric field is 30 **i** N/C and the electric potential is 100 volts.

Rank the electric potential from greatest to least at the following points within this region.

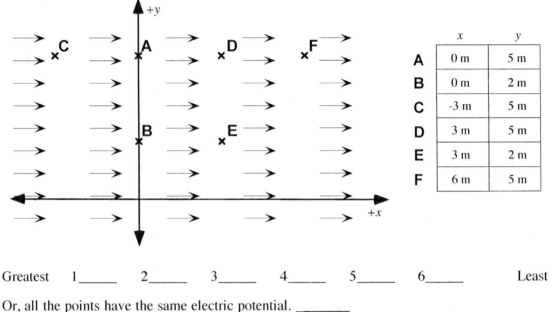

	x	y
A	0 m	5 m
B	0 m	2 m
C	-3 m	5 m
D	3 m	5 m
E	3 m	2 m
F	6 m	5 m

Greatest   1_____   2_____   3_____   4_____   5_____   6_____   Least

Or, all the points have the same electric potential. _____

Please carefully explain your reasoning.

Uniform Electric Field—Change in Potential Energy of a Positive Charge[152]

We have a large region of space that has a uniform electric field in the +x direction (⇒). At the point (0,0) m, the electric field is 30 **i** N/C and the electric potential is 100 volts.

Rank the points specified below on the basis of the change in the electric potential energy of a single positive charge of +5 C that is moved from the origin (0,0) to these particular points. That is, put first the point that will involve the largest change in electric potential energy as the charge is moved from (0,0) to that point, and put last the point that will involve the smallest change in electric potential energy as the charge is moved from (0,0) to that point. Note that some of these changes may be negative and that -5 < 2.

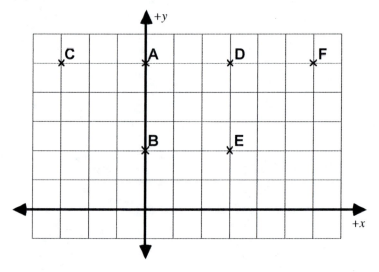

	x	y
A	0 m	5 m
B	0 m	2 m
C	-3 m	5 m
D	3 m	5 m
E	3 m	2 m
F	6 m	5 m

Largest   1_____   2_____   3_____   4_____   5_____   6_____   Smallest

Or, the +5 C charge would have the same change in electric potential energy from the origin to all of these points. _____

Or, the +5 C charge would have no change in electric potential energy from the origin to all of these points. _____

Please carefully explain your reasoning.

How sure were you of your ranking? (circle one)
Basically Guessed   Sure   Very Sure
1   2   3   4   5   6   7   8   9   10

Uniform Electric Field—Change in Potential Energy of a Negative Charge[153]

We have a large region of space that has a uniform electric field in the +x direction ($\Rightarrow$). At the point $(0,0)$ m, the electric field is 30 **i** N/C and the electric potential is 100 volts.

Rank the points specified below on the basis of the change in the electric potential energy of a single negative charge of -5 C that is moved from the origin $(0,0)$ to these particular points. That is, put first the point that will involve the largest change in electric potential energy as the charge is moved from $(0,0)$ to that point, and put last the point that will involve the smallest change in electric potential energy as the charge is moved from $(0,0)$ to that point. Note that some of these changes may be negative and that $-5 < 2$.

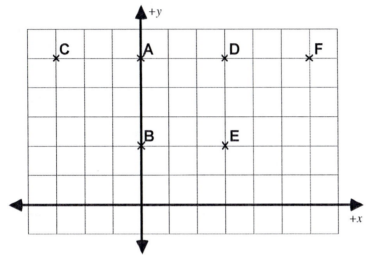

	x	y
A	0 m	5 m
B	0 m	2 m
C	-3 m	5 m
D	3 m	5 m
E	3 m	2 m
F	6 m	5 m

Largest   1____   2____   3____   4____   5____   6____   Smallest

Or, the –5 C charge would have the same change in electric potential energy from the origin to all of these points. _____

Or, the –5 C charge would have no change in electric potential energy from the origin to all of these points. _____

Please carefully explain your reasoning.

How sure were you of your ranking? (circle one)

Basically Guessed				Sure				Very Sure	
1	2	3	4	5	6	7	8	9	10

Electron Within a Charged Capacitor—Electric Potential Energy[154]

Consider an electron between the plates of a charged capacitor. The figures below show situations where the potential across the capacitor and the separations between the capacitor plates vary. Specific values are given in each figure.

Rank from greatest to least according to the magnitude of the electric potential energy of the electron within the capacitor.

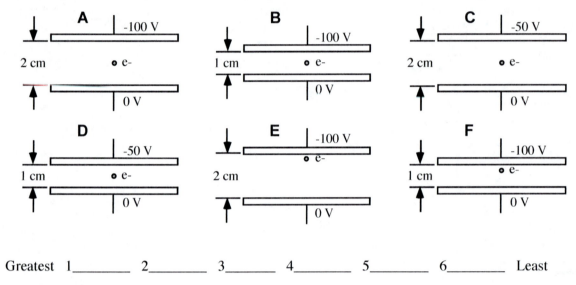

Greatest 1_____ 2_____ 3_____ 4_____ 5_____ 6_____ Least

Or, all of the electric potential energies are the same. _____

Please carefully explain your reasoning.

How sure were you of your ranking? (circle one)
Basically Guessed   Sure   Very Sure
1   2   3   4   5   6   7   8   9   10

[154] A. Van Heuvelen, S. Heath, B. Willis, L. Bryant, D. Maloney, T. O'Kuma

Basic Circuits with Identical Capacitors—Charge on a Capacitor[155]

Shown below are six capacitor circuits. All of the capacitors are identical, and all are fully charged. The batteries are also identical. In each circuit, one capacitor is labeled X.

Rank these circuits in terms of the charge on capacitor X. That is, put first the circuit in which capacitor X has the largest charge, and put last the circuit in which capacitor X has the smallest charge. If two or more circuits result in identical charges on capacitor X, give these circuits equal ranking.

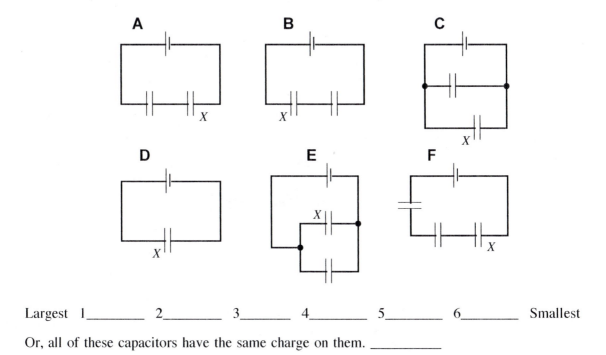

Largest 1_____ 2_____ 3_____ 4_____ 5_____ 6_____ Smallest

Or, all of these capacitors have the same charge on them. _____

Please carefully explain your reasoning.

How sure were you of your ranking? (circle one)
Basically Guessed   Sure   Very Sure
1   2   3   4   5   6   7   8   9   10

[155] L. Takahashi, D. Maloney, C. Hieggelke
Ranking Task Exercises in Physics: Student Edition   155

Simple Capacitor Circuits—Charge on a Capacitor[156]

Shown below are six capacitor circuits. All of the capacitors are either 1 μF or 2 μF, and all are fully charged. The batteries are also identical. In each circuit, one capacitor is labeled X.

Rank these circuits in terms of the charge stored in capacitor X. That is, put first the circuit in which capacitor X has the largest charge stored, and put last the circuit in which capacitor X has the smallest charge stored. If two or more circuits result in identical charge stored for capacitor X, give these circuits equal ranking.

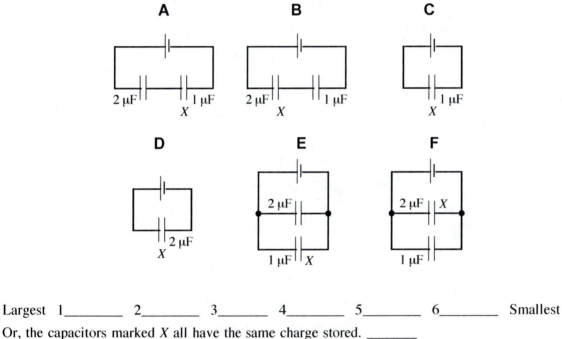

Largest 1_____ 2_____ 3_____ 4_____ 5_____ 6_____ Smallest

Or, the capacitors marked X all have the same charge stored. _____

Please carefully explain your reasoning.

How sure were you of your ranking? (circle one)

Basically Guessed				Sure				Very Sure	
1	2	3	4	5	6	7	8	9	10

[156] L. Takahashi, C. Hieggelke

Simple Capacitor Circuits—Charge on Capacitors[157]

Shown below are six capacitors in two circuits. The capacitors are 1 μF, 2 μF, or 3 μF and all are fully charged. The batteries in the two circuits are ideal and identical.

Rank these capacitors in terms of the charge stored in each capacitor. That is, put first the capacitor that has the largest charge stored, and put last the capacitor that has the smallest charge stored. If two or more capacitors result in identical charge stored, give these capacitors equal ranking.

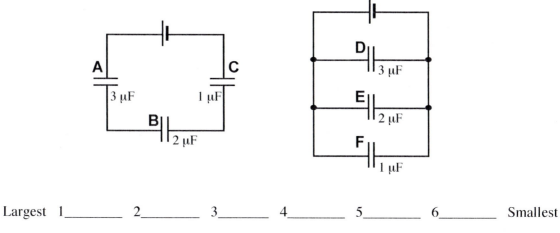

Largest 1_____ 2_____ 3_____ 4_____ 5_____ 6_____ Smallest

Or, all six capacitors have the same charge stored. _____

Please carefully explain your reasoning.

How sure were you of your ranking? (circle one)

Basically Guessed				Sure				Very Sure	
1	2	3	4	5	6	7	8	9	10

Basic Circuits with Identical Capacitors—Voltage Across a Capacitor[158]

Shown below are six capacitor circuits. All of the capacitors are identical and all are fully charged. The batteries are also identical. In each circuit, one capacitor is labeled X.

Rank these circuits in terms of the voltage across capacitor X. That is, put first the circuit in which capacitor X has the largest voltage, and put last the circuit in which capacitor X has the smallest voltage. If two or more circuits result in identical voltages for capacitor X, give these circuits equal ranking.

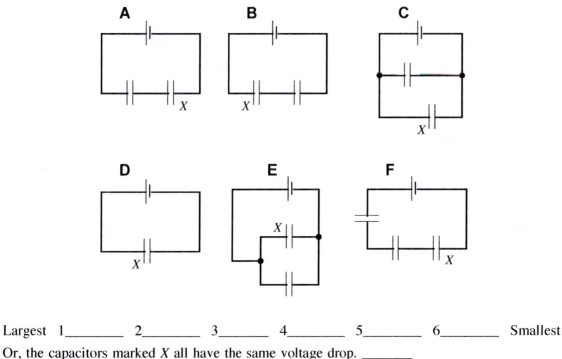

Largest 1_____ 2_____ 3_____ 4_____ 5_____ 6_____ Smallest

Or, the capacitors marked X all have the same voltage drop. _____

Please carefully explain your reasoning.

How sure were you of your ranking? (circle one)
Basically Guessed   Sure   Very Sure
1   2   3   4   5   6   7   8   9   10

[158] L. Takahashi, C. Hieggelke, D. Maloney
Ranking Task Exercises in Physics: Student Edition   158

Circuits with Resistors and Capacitors—Current[159]

The circuit shown below has circuit elements α, β, γ, and δ, which are either a 1 F capacitor or a 1, 2, or 4 ohm resistor.

Rank the arrangements of circuit elements given below in order from largest to smallest magnitude of current that flows after a long time through the 6 volt battery connected as shown.

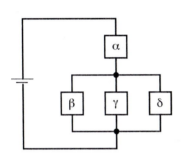

Arrangement	α	β	γ	δ
A	1F = 1 Farad	4Ω = 4 Ohms	1Ω = 1 Ohm	2Ω = 2 Ohms
B	1Ω	1F	2Ω	4Ω
C	2Ω	1Ω	4Ω	1F
D	4Ω	2Ω	1F	1Ω
E	1F	1Ω	2Ω	4Ω
F	1Ω	4Ω	2Ω	1F

Largest 1_____ 2_____ 3_____ 4_____ 5_____ 6_____ Smallest

Or, the magnitudes of the currents are the same for all six arrangements. _____

Please carefully explain your reasoning.

How sure were you of your ranking? (circle one)

Basically Guessed   Sure   Very Sure
1   2   3   4   5   6   7   8   9   10

[159] K. Wetz

Current Carrying Wires with Different Lengths—Net Charge[160]

The figures below show six segments of wires that are carrying electric currents. In all six cases, the currents are flowing to the right. As you can see from the values in the figures, the pieces of the wires shown have different lengths, and they are carrying different currents. For the ranking below, we are only interested in the segments of the wires actually shown in the figures.

Rank these segments from greatest to smallest on the basis of the net electric charge, i.e., the difference in number of positive charges and negative charges in each segment. That is, put first the segment that has the greatest net electric charge on it, and put last the segment that has the smallest net electric charge on it.

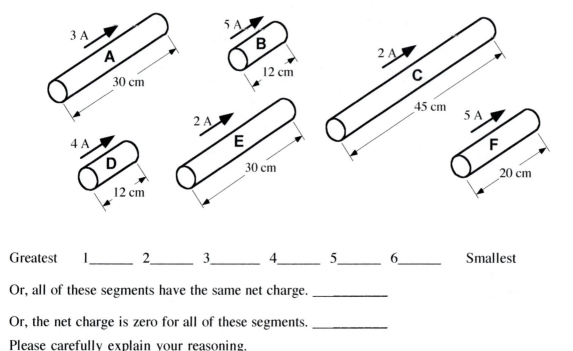

Greatest   1_____ 2_____ 3_____ 4_____ 5_____ 6_____   Smallest

Or, all of these segments have the same net charge. _____

Or, the net charge is zero for all of these segments. _____

Please carefully explain your reasoning.

How sure were you of your ranking? (circle one)
Basically Guessed   Sure   Very Sure
1   2   3   4   5   6   7   8   9   10

Batteries and Bulbs—Bulb Brightness[161]

Identical batteries are connected in different arrangements to the identical light bulbs.

Rank from brightest to least bright these items on the basis of bulb brightness.

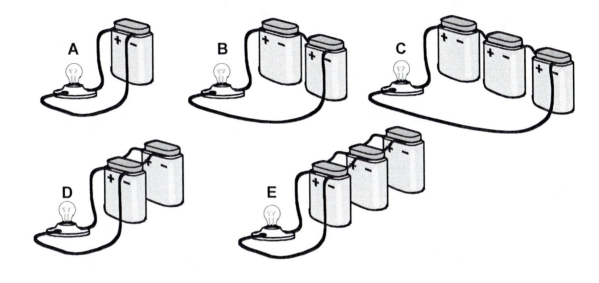

Brightest 1_____ 2_____ 3_____ 4_____ 5_____ Least Bright

Or, all bulbs have the same brightness. _____

Please carefully explain your reasoning.

How sure were you of your ranking? (circle one)
Basically Guessed   Sure   Very Sure
1   2   3   4   5   6   7   8   9   10

[161] C. Hieggelke, D. Maloney, T. O'Kuma
Ranking Task Exercises in Physics: Student Edition   161

Simple Resistor Circuits—Resistance[162]

All the resistors in the circuits below are identical. The batteries are ideal with voltages of 12 volts. All connecting wires have negligible resistance.

Rank these on how hard it will be for the battery to drive current through each circuit, from greatest to least. Circle any ties.

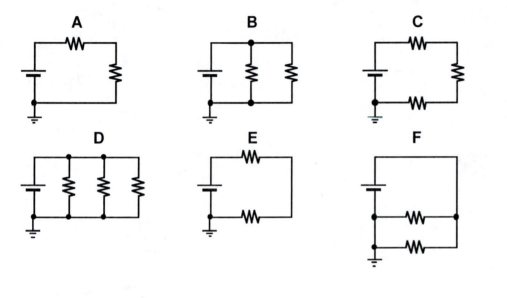

Greatest 1_____ 2_____ 3_____ 4_____ 5_____ 6_____ Least

Or, the resistance is the same in each situation. _____

Please carefully explain your reasoning.

How sure were you of your ranking?

Basically guessed				Sure				Very Sure	
1	2	3	4	5	6	7	8	9	10

Simple Resistor Circuits—Current I[163]

All the resistors in the circuits below are identical. The batteries are ideal with voltages of 6 or 12 volts as shown below. All connecting wires have negligible resistance.

Rank the current passing through the upper right hand corner of each circuit, from greatest to least. Circle any ties.

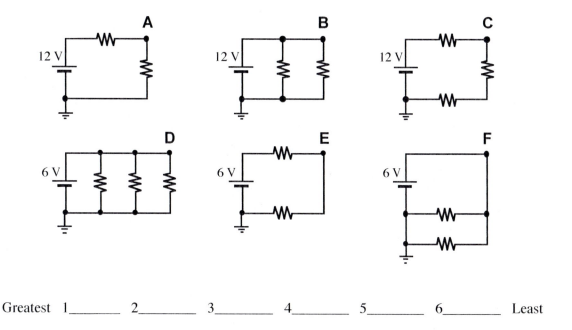

Greatest 1_____ 2_____ 3_____ 4_____ 5_____ 6_____ Least

Or, the current is the same in each situation. _____

Please carefully explain your reasoning.

How sure were you of your ranking? (circle one)
Basically Guessed   Sure   Very Sure
1   2   3   4   5   6   7   8   9   10

Simple Resistor Circuits—Voltage[164]

All the resistors in the circuits below are identical. The batteries are ideal with voltages of 6 or 12 volts as shown below. All connecting wires have negligible resistance.

Rank the voltage at the upper right hand corner, relative to the ground. Place the highest voltage first and the lowest voltage last. Circle any ties.

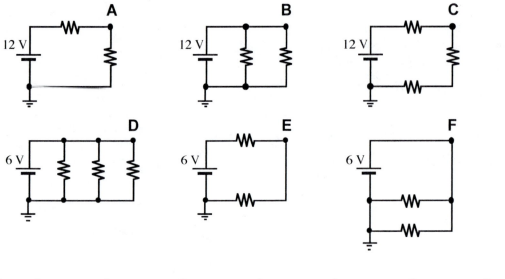

Highest 1_____ 2_____ 3_____ 4_____ 5_____ 6_____ Lowest

Or, the voltage is the same in each situation. _____

Please carefully explain your reasoning.

How sure were you of your ranking? (circle one)
Basically Guessed   Sure   Very Sure
1   2   3   4   5   6   7   8   9   10

[164] T. O'Kuma
Ranking Task Exercises in Physics: Student Edition   164

Simple Resistor Circuits—Current II[165]

All the resistors in the circuits below are identical at 2 Ω each. The batteries are ideal with voltages of 6 or 12 volts as shown below. All connecting wires have negligible resistance.

Rank the current passing through the upper right hand corner of each circuit from greatest to least.

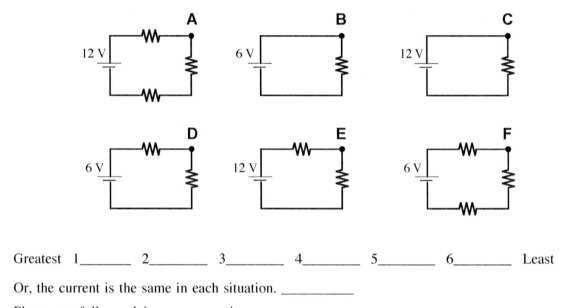

Greatest 1_____ 2_____ 3_____ 4_____ 5_____ 6_____ Least

Or, the current is the same in each situation. _____

Please carefully explain your reasoning.

How sure were you of your ranking? (circle one)

Basically Guessed				Sure				Very Sure	
1	2	3	4	5	6	7	8	9	10

[165] C. Hill, C. Lam, M. Mann

Simple Circuits—Bulb Brightness[166]

Rank the following according to the brightness of the bulb labeled *X* from brightest to dimmest. All the bulbs are identical and you should treat them as ohmic resistors. The batteries are ideal (no internal resistance) and you should treat the wires as zero resistance connectors.

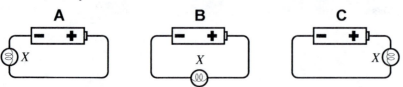

1. brightest _____ _____ _____ dimmest ___ Confidence Level (1 low-10 high)

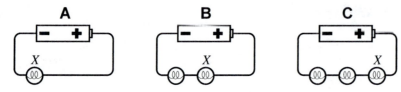

2. brightest _____ _____ _____ dimmest ___ Confidence Level (1 low-10 high)

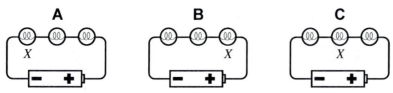

3. brightest _____ _____ _____ dimmest ___ Confidence Level (1 low-10 high)

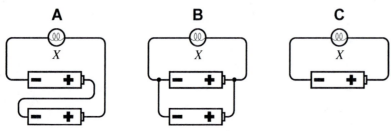

4. brightest _____ _____ _____ dimmest ___ Confidence Level (1 low-10 high)

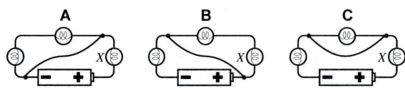

5. brightest _____ _____ _____ dimmest ___ Confidence Level (1 low-10 high)

Please carefully explain your reasoning.

Simple Circuits—Ammeter Reading[167]

Rank according to the ammeter reading (current) from largest to smallest. Assume that the ammeter has zero resistance. All the bulbs are identical and you should treat them as ohmic resistors. The batteries are ideal (no internal resistance) and you should treat the wires as zero resistance connectors.

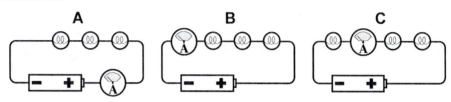

1. largest _____ _____ _____ smallest _____ Confidence Level (1 low-10 high)

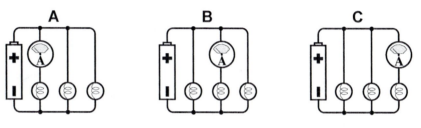

2. largest _____ _____ _____ smallest _____ Confidence Level (1 low-10 high)

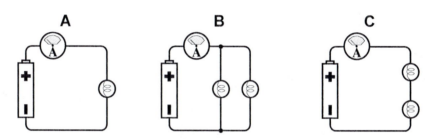

3. largest _____ _____ _____ smallest _____ Confidence Level (1 low-10 high)

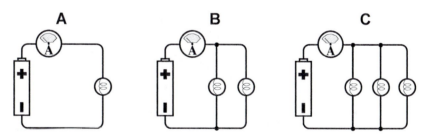

4. largest _____ _____ _____ smallest _____ Confidence Level (1 low-10 high)

Please carefully explain your reasoning.

Simple Circuits—Potential Difference I[168]

Rank according to magnitude of difference in potential between points N and M from largest to smallest. All the bulbs are identical and you should treat them as ohmic resistors. The batteries are ideal (no internal resistance) and you should treat the wires as zero resistance connectors.

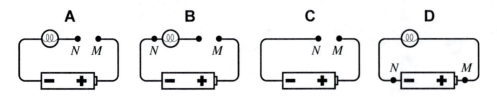

1. largest _____ _____ _____ _____ smallest ___ Confidence Level (1 low-10 high)

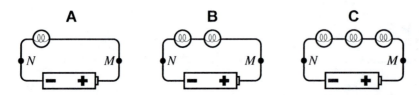

2. largest _____ _____ _____ smallest ___ Confidence Level (1 low-10 high)

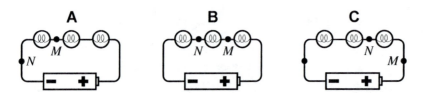

3. largest _____ _____ _____ smallest ___ Confidence Level (1 low-10 high)

Please carefully explain your reasoning.

Simple Circuits—Potential Difference II[169]

Rank according to magnitude of difference in potential between points N and M from largest to smallest. All the bulbs are identical and you should treat them as ohmic resistors. The batteries are ideal (no internal resistance) and you should treat the wires as zero resistance connectors.

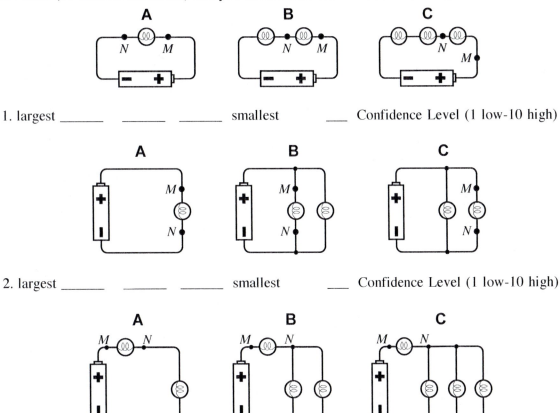

1. largest _____ _____ _____ smallest ___ Confidence Level (1 low-10 high)

2. largest _____ _____ _____ smallest ___ Confidence Level (1 low-10 high)

3. largest _____ _____ _____ smallest ___ Confidence Level (1 low-10 high)

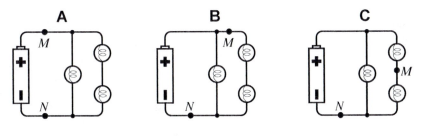

4. largest _____ _____ _____ smallest ___ Confidence Level (1 low-10 high)

Please carefully explain your reasoning.

Simple Circuits—Equivalent Resistor[170]

Rank according to the size of single resistor that would be equivalent to the given group of resistors from largest to smallest. All the bulbs are identical and you should treat them as ohmic resistors. The batteries are ideal (no internal resistance) and you should treat the wires as zero resistance connectors.

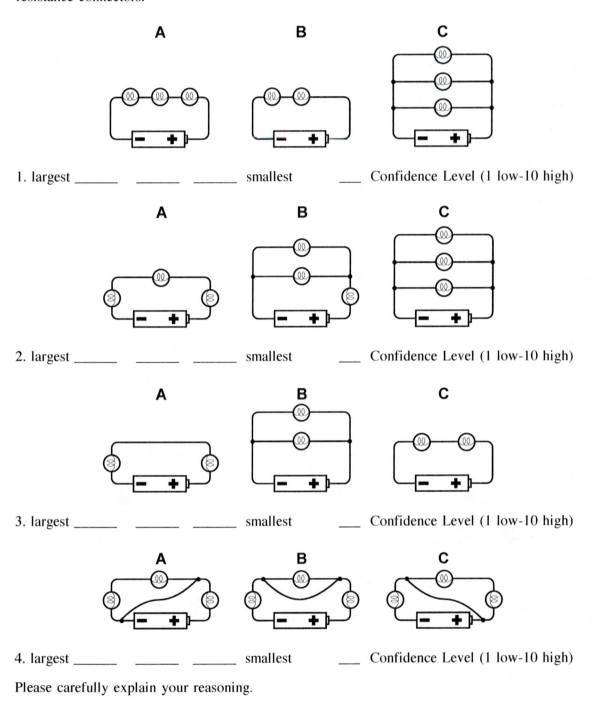

1. largest _____   _____   _____ smallest   ___ Confidence Level (1 low-10 high)

2. largest _____   _____   _____ smallest   ___ Confidence Level (1 low-10 high)

3. largest _____   _____   _____ smallest   ___ Confidence Level (1 low-10 high)

4. largest _____   _____   _____ smallest   ___ Confidence Level (1 low-10 high)

Please carefully explain your reasoning.

[170] D. Albers

Current Carrying Wires with Different Resistances—Net Charge[171]

Shown below are six segments of wires that are carrying electric currents. All of these segments have the same length and the same diameter, but the wires are made of different materials so they have different resistances. The currents flowing in these segments also vary. Specific values for each of these properties are given in the figures.

Rank these situations from greatest to smallest on the basis of the net electric charge on each segment, i.e., on the basis of the difference between the number of positive and negative charges in the wire segment. Put first the situation that has the greatest net charge, and put last the situation with the smallest net charge, i.e., the smallest difference between the number of positive and negative charges.

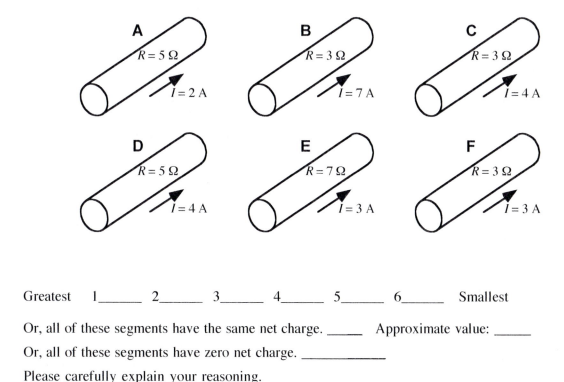

Greatest   1_____ 2_____ 3_____ 4_____ 5_____ 6_____   Smallest

Or, all of these segments have the same net charge. _____   Approximate value: _____

Or, all of these segments have zero net charge. _____

Please carefully explain your reasoning.

How sure were you of your ranking? (circle one)
Basically Guessed   Sure   Very Sure
1   2   3   4   5   6   7   8   9   10

Simple Resistor Circuits—Voltage Drop[172]

All the resistors in the circuits below are identical at 2 Ω each. The batteries are ideal with voltages of 6 or 12 volts as shown. All connecting wires have negligible resistance.

Rank the voltage at the upper right-hand corner relative to the ground. Place the highest voltage first and the lowest voltage last.

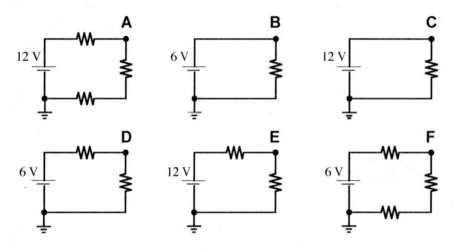

Highest 1_____ 2_____ 3_____ 4_____ 5_____ 6_____ Lowest

Or, all of the voltages are the same. _____

Please carefully explain your reasoning.

How sure were you of your ranking? (circle one)

Basically Guessed					Sure				Very Sure	
1	2	3	4	5	6	7	8	9	10

All resistors shown are identical in the circuits below. The switches in each case are open and remain open.

Rank the voltmeter readings from highest to lowest.

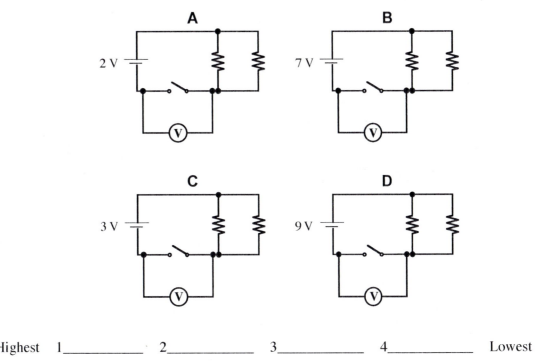

Highest   1_____   2_____   3_____   4_____   Lowest

Or, all of the readings are zero. _____

Please carefully explain your reasoning.

How sure were you of your ranking? (circle one)
Basically Guessed   Sure   Very Sure
1   2   3   4   5   6   7   8   9   10

All resistors shown are identical in the circuits below. The switches in each case are open and remain open.

Rank the voltmeter readings from highest to lowest.

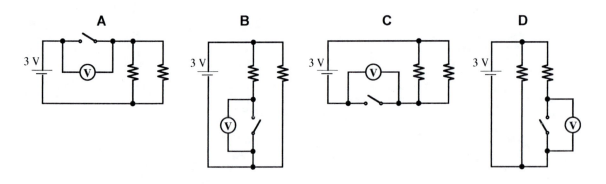

Highest   1_____   2_____   3_____   4_____   Lowest

Or, all readings are zero. _____

Or, all readings are 3 V. _____

Or, not enough information is given to be able to rank these readings. _____

Please carefully explain your reasoning.

How sure were you of your ranking? (circle one)
Basically Guessed   Sure   Very Sure
1   2   3   4   5   6   7   8   9   10

Circuit with Two Open and Closed Switches—Ammeter Readings[175]

Shown below is a DC circuit that contains two switches. The resistance of each switch is negligible when closed. All of the connecting wires should be considered to have zero resistance. All of the resistors shown are identical. The circuit contains an ideal ammeter and an ideal voltmeter. The diagram shows the switches open. Beside the diagram are four different switch configurations for the circuit.

Rank these configurations from largest to smallest in terms of the ammeter reading.

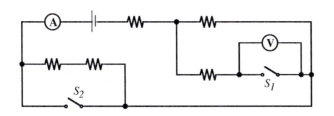

Configuration	S_1	S_2
A	Open	Open
B	Open	Closed
C	Closed	Open
D	Closed	Closed

Largest   1_____   2_____   3_____   4_____   Smallest

Or, all configurations produce the same ammeter reading. _____

Or, all configurations produce a zero ammeter reading. _____

Please carefully explain your reasoning.

How sure were you of your ranking? (circle one)

Basically Guessed   Sure   Very Sure

1   2   3   4   5   6   7   8   9   10

Circuit with Two Open and Closed Switches—Voltmeter Readings[176]

Shown below is a DC circuit that contains two switches. The resistance of each switch is negligible when closed. All of the connecting wires should be considered to have zero resistance. All of the resistors shown are identical. The circuit contains an ideal ammeter and an ideal voltmeter. The diagram shows the switches open. Beside the diagram are four different switch configurations for the circuit.

Rank these configurations from largest to smallest in terms of the voltmeter reading.

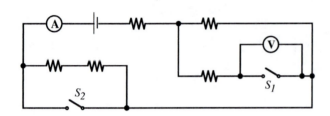

Configuration	S_1	S_2
A	Open	Open
B	Open	Closed
C	Closed	Open
D	Closed	Closed

Largest   1_____   2_____   3_____   4_____   Smallest

Or, all configurations produce the same voltmeter reading. _____

Or, all configurations produce a zero voltmeter reading. _____

Please carefully explain your reasoning.

How sure were you of your ranking? (circle one)

Basically Guessed   Sure   Very Sure
1   2   3   4   5   6   7   8   9   10

[176] L. Takahashi, C. Hieggelke

Shown below is a DC circuit that contains a number of switches. The resistance of each switch is negligible when closed. All of the connecting wires may be considered to have zero resistance. All of the resistors shown are identical. The circuit contains an ideal voltmeter. The diagram shows all of the switches open. Beside the diagram are eight different switch configurations for the circuit.

Rank these configurations in terms of the voltmeter reading. That is, put first the configuration for which the voltmeter gives the largest reading, and put last the configuration for which the voltmeter gives the smallest reading. If two or more configurations produce the same voltmeter reading, give these configurations equal ranking.

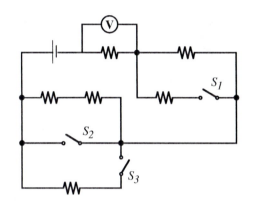

Configuration	S_1	S_2	S_3
A	Open	Open	Open
B	Open	Open	Closed
C	Open	Closed	Open
D	Open	Closed	Closed
E	Closed	Open	Open
F	Closed	Open	Closed
G	Closed	Closed	Open
H	Closed	Closed	Closed

Largest   1_____   2_____   3_____   4_____   5_____   6_____   7_____   8_____   Smallest

Or, all configurations produce the same voltmeter reading. _____

Please carefully explain your reasoning.

How sure were you of your ranking? (circle one)

Basically Guessed   Sure   Very Sure

1   2   3   4   5   6   7   8   9   10

Circuit with Three Open and Closed Switches—Voltmeter Readings II[178]

Shown below is a DC circuit that contains a number of switches. The resistance of each switch is negligible when closed. All of the connecting wires may be considered to have zero resistance. All of the resistors shown are identical. The circuit contains an ideal voltmeter. The diagram shows all of the switches open. Beside the diagram are eight different switch configurations for the circuit.

Rank these configurations in terms of the voltmeter reading. That is, put first the configuration for which the voltmeter gives the largest reading, and put last the configuration for which the voltmeter gives the smallest reading. If two or more configurations produce the same voltmeter reading, give these configurations equal ranking.

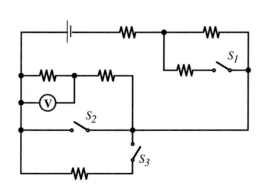

Configuration	S_1	S_2	S_3
A	Open	Open	Open
B	Open	Open	Closed
C	Open	Closed	Open
D	Open	Closed	Closed
E	Closed	Open	Open
F	Closed	Open	Closed
G	Closed	Closed	Open
H	Closed	Closed	Closed

Largest   1_____ 2_____ 3_____ 4_____ 5_____ 6_____ 7_____ 8_____   Smallest

Or, all configurations produce the same voltmeter reading. _____

Please carefully explain your reasoning.

How sure were you of your ranking? (circle one)
Basically Guessed   Sure   Very Sure
1   2   3   4   5   6   7   8   9   10

Circuit with Three Open and Closed Switches—Ammeter Readings[179]

Shown below is a DC circuit that contains a number of switches. The resistance of each switch is negligible when closed. All of the connecting wires may be considered to have zero resistance. All of the resistors shown are identical. The circuit contains an ideal ammeter. The diagram shows all of the switches open. Beside the diagram are eight different switch configurations for the circuit.

Rank these configurations in terms of the ammeter reading. That is, put first the configuration for which the ammeter gives the largest reading, and put last the configuration for which the ammeter gives the smallest reading. If two or more configurations produce the same ammeter reading, give these configurations equal ranking.

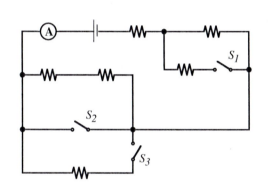

Configuration	S_1	S_2	S_3
A	Open	Open	Open
B	Open	Open	Closed
C	Open	Closed	Open
D	Open	Closed	Closed
E	Closed	Open	Open
F	Closed	Open	Closed
G	Closed	Closed	Open
H	Closed	Closed	Closed

Largest   1_____   2_____   3_____   4_____   5_____   6_____   7_____   8_____   Smallest

Or, all configurations produce the same ammeter reading. _____

Please carefully explain your reasoning.

How sure were you of your ranking? (circle one)
Basically Guessed   Sure   Very Sure
1   2   3   4   5   6   7   8   9   10

[179] L. Takahashi

Moving Charges in Uniform Magnetic Field—Acceleration[180]

Moving charged particles are released with a velocity (details listed below) at the point P (2 m, 2 m) in a large region of space which has a uniform magnetic field in the $+x$ direction. All these particles have the same mass, and they are released individually into this field.

Rank from greatest to least the magnitude of the initial acceleration of these charged particles as they are released from P.

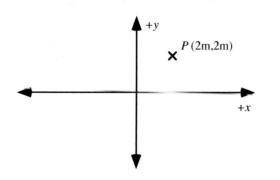

Case	Charge	Speed	Direction
A	+5 mC	3 m/s	$+x$
B	+5 mC	3 m/s	$-x$
C	+5 mC	3 m/s	$+y$
D	+5 mC	3 m/s	$-y$
E	-10 mC	5 m/s	$+y$
F	+10 mC	5 m/s	$-y$

Greatest 1_____ 2_____ 3_____ 4_____ 5_____ 6_____ Least

Or, the magnitude of the initial acceleration will be the same for all of these (but not zero). _____

Or, the magnitude of the initial acceleration will be zero for all of these. _____

Please carefully explain your reasoning.

Moving Charges in Uniform Magnetic Field—Change in Kinetic Energy[181]

Moving charged particles are released with a velocity (details listed below) at the point P (2 m, 2 m) in a large region of space that has a uniform magnetic field in the $+x$ direction. All these particles have the same mass, and they are released individually into this field.

Rank from greatest to least the change in kinetic energy of these charged particles after they have traveled 1 cm from P.

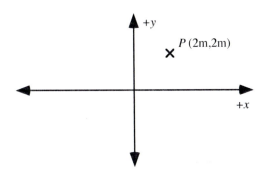

Case	Charge	Speed	Direction
A	+5 mC	3 m/s	$+x$
B	+5 mC	3 m/s	$-x$
C	+5 mC	3 m/s	$+y$
D	+5 mC	3 m/s	$-y$
E	-10 mC	5 m/s	$+y$
F	+10 mC	5 m/s	$-y$

Greatest 1_____ 2_____ 3_____ 4_____ 5_____ 6_____ Least

Or, the change in kinetic energy will be the same for all of these (but not zero). _____

Or, the change in kinetic energy will be zero for all of these. _____

Please carefully explain your reasoning.

Currents at Corners of Squares—Magnetic Field at Center[182]

Shown below are six situations where electric current carrying wires are at the corners of squares. We are viewing the situations from "end-on" of the wires. All of the currents have the same magnitude, but they vary in whether they flow into or out of the page.

Rank these situations, from greatest to least, on the basis of the net magnetic fields at the centers of the squares.

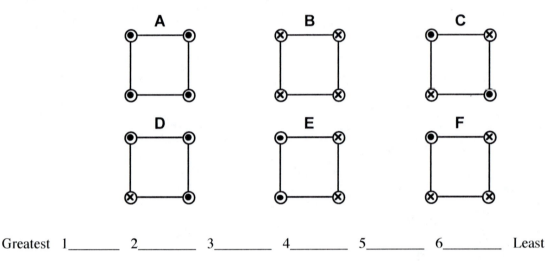

Greatest 1_____ 2_____ 3_____ 4_____ 5_____ 6_____ Least

Or, the magnetic field at the center will be the same for all six arrangements. _____

Or, the magnetic field at the center will be zero for all six arrangements._____

Please carefully explain your reasoning.

How sure were you of your ranking? (circle one)
Basically Guessed   Sure   Very Sure
1   2   3   4   5   6   7   8   9   10

[182] D. Maloney, C. Hieggelke

Magnetic Fields by Currents at Corners of Squares—Magnetic Field at Center[183]

Shown below are six situations where electric current carrying wires are at the corners of squares. We are viewing the situations from "end-on" of the wires. All of the currents have the same magnitude, but they vary in whether they flow into or out of the page. Also shown in each figure are the directions of the net magnetic fields produced by the currents at the midpoints of the four sides of the squares.

Rank these situations, from greatest to least, on the basis of the net magnetic fields at the centers of the squares.

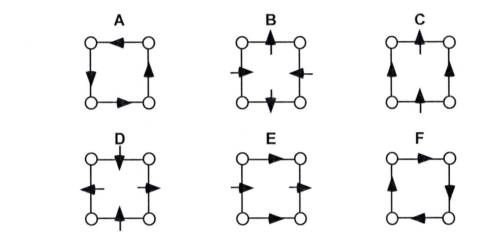

Greatest 1_____ 2_____ 3_____ 4_____ 5_____ 6_____ Least

Or, the magnetic field at the center will be the same for all six arrangements. _____

Or, the magnetic field at the center will be zero for all six arrangements. _____

Please carefully explain your reasoning.

How sure were you of your ranking? (circle one)

Basically Guessed				Sure				Very Sure	
1	2	3	4	5	6	7	8	9	10

Bars Moving in Magnetic Fields—Force on Electrons[184]

Shown below are three bars moving at different speeds through regions in which there are uniform magnetic fields. All of the bars are made of the same conducting material and all are moving in a plane perpendicular to the same magnetic field. The bars have various lengths (L, $.75L$, and $.25L$) and are moving at three constant but different speeds. Within each bar two electrons, one at the end and the other at the middle, are pictured and labeled.

Rank these electrons, from greatest to least, on the basis of the strength (magnitude) of the magnetic forces on these electrons.

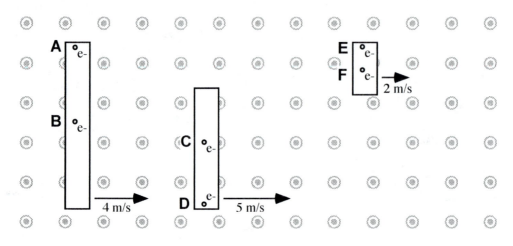

Greatest 1_____ 2_____ 3_____ 4_____ 5_____ 6_____ Least

Or, all six electrons experience the same non-zero magnetic force. _____

Or, all six electrons experience no magnetic force._____

Please carefully explain your reasoning.

How sure were you of your ranking? (circle one)

Basically Guessed				Sure					Very Sure
1	2	3	4	5	6	7	8	9	10

[184] D. Maloney, C. Hieggelke

Charged Particle and a Uniform Magnetic Field—Acceleration[185]

The figures below show the circular paths of six charged particles that have been injected into six different but uniform magnetic fields. The particles have the same mass, and they were all given the same initial speed before they entered the field. However, the charges on the particles and the radii of their paths vary.

Rank these situations from greatest to least on the basis of the magnitude of the acceleration that each charge is experiencing.

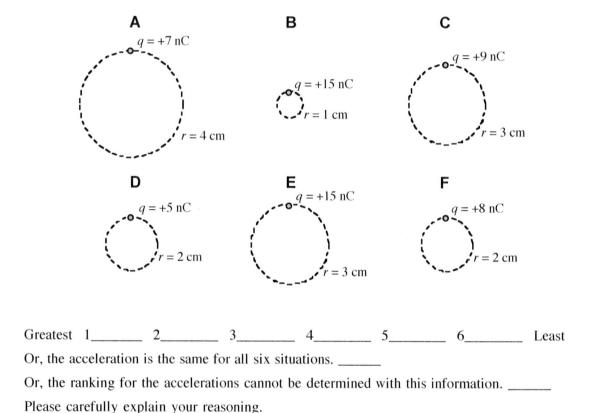

Greatest 1_____ 2_____ 3_____ 4_____ 5_____ 6_____ Least

Or, the acceleration is the same for all six situations. _____

Or, the ranking for the accelerations cannot be determined with this information. _____

Please carefully explain your reasoning.

How sure were you of your ranking? (circle one)

Basically Guessed				Sure				Very Sure	
1	2	3	4	5	6	7	8	9	10

[185] C. Hieggelke, D. Maloney, T. O'Kuma
Ranking Task Exercises in Physics: Student Edition   185

Charged Particle and a Uniform Magnetic Field—Force[186]

The figures below show the circular paths of six charged particles that have been injected into six different but uniform magnetic fields. The particles have the same mass, and they were all given the same initial speed before they entered the field. However, the charges on the particles and the radii of their paths vary.

Rank these situations from greatest to least on the basis of the magnitude of the force on each charge.

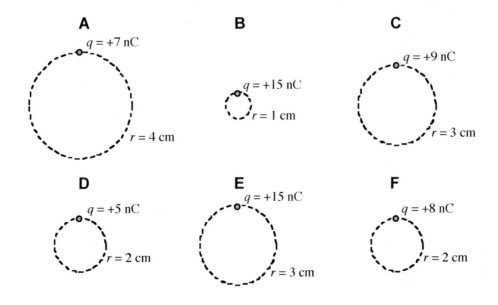

Greatest 1_____ 2_____ 3_____ 4_____ 5_____ 6_____ Least

Or, the force is the same for all six situations. _____

Or, there is no force on any of these six particles. _____

Or, the ranking for the forces cannot be determined with this information. _____

Please carefully explain your reasoning.

How sure were you of your ranking? (circle one)
Basically Guessed   Sure   Very Sure
1   2   3   4   5   6   7   8   9   10

Charged Particles in Uniform Magnetic Fields—Work Done[187]

The figures below show the circular paths of six charged particles that have been injected into six different but uniform magnetic fields. The particles have the same mass, and they were all given the same initial speed before they entered the field. However, the charges on the particles and the radii of their paths vary.

Rank these situations from greatest to least on the basis of the work done on each charge during one complete rotation.

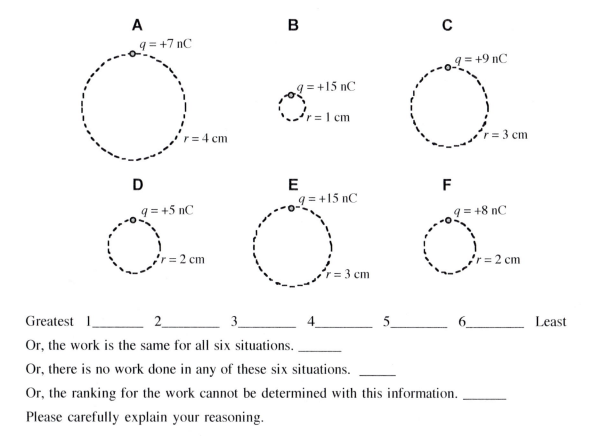

Greatest 1_____ 2_____ 3_____ 4_____ 5_____ 6_____ Least

Or, the work is the same for all six situations. _____

Or, there is no work done in any of these six situations. _____

Or, the ranking for the work cannot be determined with this information. _____

Please carefully explain your reasoning.

How sure were you of your ranking? (circle one)

Basically Guessed				Sure				Very Sure	
1	2	3	4	5	6	7	8	9	10

[187] D. Maloney, C. Hieggelke

Charged Particles in Uniform Magnetic Fields—Magnetic Field[188]

The figures below show the circular paths of six charged particles that have been injected into six different but uniform magnetic fields. The particles have the same mass, and they were all given the same initial speed before they entered the field. However, the charges on the particles and the radii of their paths vary.

Rank these situations from greatest to least on the basis of the magnitude of the magnetic field on each charge.

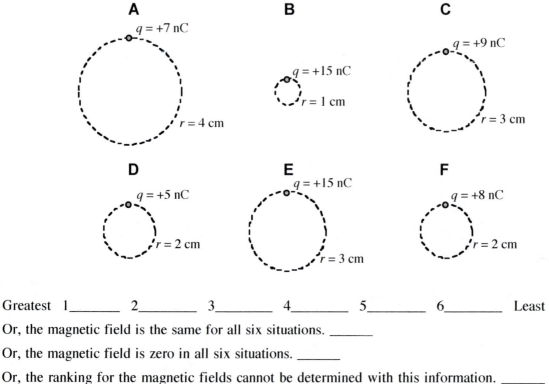

A
$q = +7$ nC
$r = 4$ cm

B
$q = +15$ nC
$r = 1$ cm

C
$q = +9$ nC
$r = 3$ cm

D
$q = +5$ nC
$r = 2$ cm

E
$q = +15$ nC
$r = 3$ cm

F
$q = +8$ nC
$r = 2$ cm

Greatest 1_____ 2_____ 3_____ 4_____ 5_____ 6_____ Least

Or, the magnetic field is the same for all six situations. _____

Or, the magnetic field is zero in all six situations. _____

Or, the ranking for the magnetic fields cannot be determined with this information. _____

Please carefully explain your reasoning.

How sure were you of your ranking? (circle one)

Basically Guessed					Sure				Very Sure	
1	2	3	4	5	6	7	8	9	10

[188] C. Hieggelke, D. Maloney

Charges Near Magnets—Magnetic Force[189]

The figures below show electrically charged particles sitting at rest near the poles of permanent magnets. All of the permanent magnets are the same strength. The magnitudes and signs of the electric charges are shown in the circles, which represent the particles. The zero values indicate that the particle is actually electrically neutral.

Rank these situations, from strongest attraction to strongest repulsion, on the basis of the magnetic force exerted by the magnet on the charge. As stated, if the interaction is actually a repulsion rather than an attraction, it would be ranked lower than any attraction.

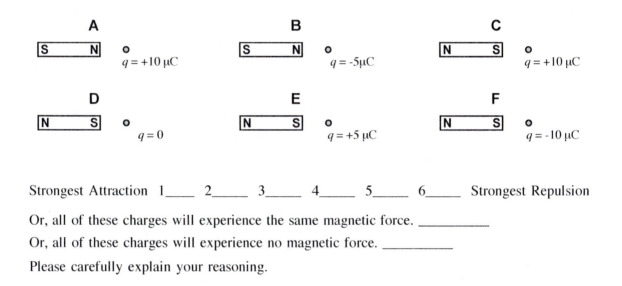

Strongest Attraction 1____ 2_____ 3_____ 4_____ 5_____ 6_____ Strongest Repulsion

Or, all of these charges will experience the same magnetic force. _____

Or, all of these charges will experience no magnetic force. _____

Please carefully explain your reasoning.

How sure were you of your ranking? (circle one)

Basically Guessed				Sure				Very Sure	
1	2	3	4	5	6	7	8	9	10

Pairs of Long Current Carrying Wires—Magnetic Field[190]

Shown below are several situations with two wires with currents of 2 A flowing out of (•) or into (X) the page.

Rank the magnetic field at point P from greatest to least. Count a downward pointing field as negative and upward as positive. You may assume that each wire or point of interest are separated by the same distance from the adjacent items and that each situation is independent of all others.

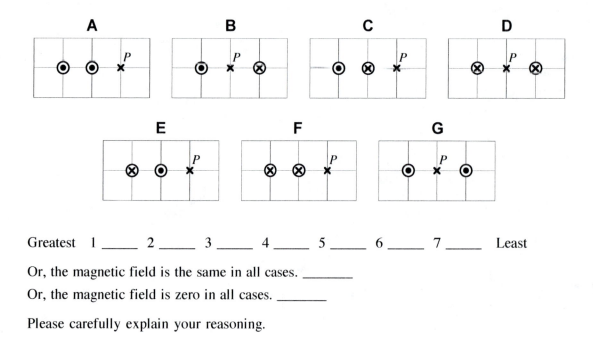

Greatest 1 _____ 2 _____ 3 _____ 4 _____ 5 _____ 6 _____ 7 _____ Least

Or, the magnetic field is the same in all cases. _____

Or, the magnetic field is zero in all cases. _____

Please carefully explain your reasoning.

How sure were you of your ranking? (Circle one)
Basically Guessed   Sure   Very Sure
1   2   3   4   5   6   7   8   9   10

[190] C. Davies, B. Diesslin, M. Nelson, G. Shepherd
Ranking Task Exercises in Physics: Student Edition   190

Pairs of Equal Current Electromagnets—Force[191]

Six pairs of electromagnets are shown below. The same current flows in each case. The pairs of electromagnets are separated by the same distance, and they have the same length and diameter. Carefully observe the orientation of the coil and direction of current flow.

Rank the forces between the magnets from the most attractive to the most repulsive.

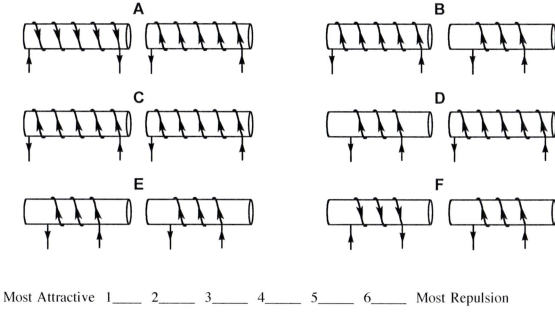

Most Attractive 1____ 2_____ 3_____ 4_____ 5_____ 6_____ Most Repulsion

Or, all of these pairs exert equal forces on each other. _____

Please carefully explain your reasoning.

How sure were you of your ranking? (circle one)
Basically Guessed   Sure   Very Sure
1   2   3   4   5   6   7   8   9   10

Pairs of Equal Current Electromagnets—Magnetic Field Between[192]

Six pairs of electromagnets are shown below. The same current flows in each electromagnet. The pairs are separated by the same distance, and they have the same length and diameter. Carefully observe the orientation of the coil and direction of current flow in the electromagnets.

Rank the magnitudes of the magnetic fields at the midpoint between the electromagnets from the largest to the smallest.

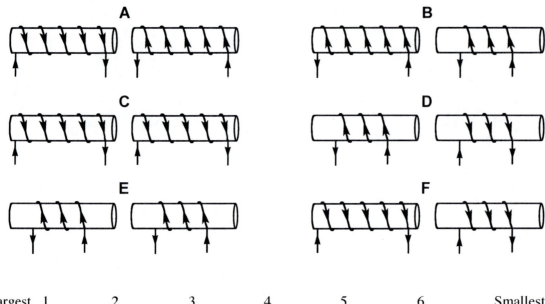

Largest 1_____ 2_____ 3_____ 4_____ 5_____ 6_____ Smallest

Or, the magnitudes of the magnetic field are all the same._____

Please carefully explain your reasoning.

How sure were you of your ranking? (circle one)

Basically Guessed   Sure   Very Sure
1   2   3   4   5   6   7   8   9   10

Unequal Current Electromagnets—Magnetic Field at Ends[193]

Six electromagnets are shown below. The current in each electromagnet is indicated in the diagram as well as the number of turns. They have the same length and diameter. Carefully observe the orientation of the turns and direction of current flow in the coil.

Rank the magnetic field just outside the right end of the electromagnet from the largest to the smallest. (The direction to the right is positive, and –5 < -3.)

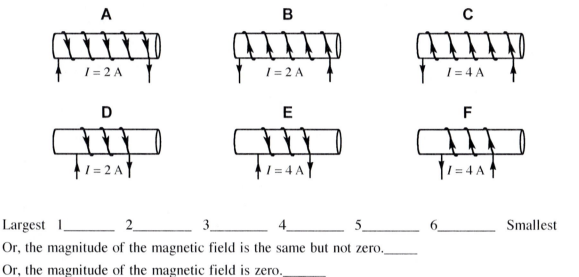

Largest 1_____ 2_____ 3_____ 4_____ 5_____ 6_____ Smallest

Or, the magnitude of the magnetic field is the same but not zero._____

Or, the magnitude of the magnetic field is zero._____

Please carefully explain your reasoning.

How sure were you of your ranking? (circle one)

Basically Guessed   Sure   Very Sure

1   2   3   4   5   6   7   8   9   10

[193] B. Emerson, C. Hieggelke
Ranking Task Exercises in Physics: Student Edition   193

Electromagnets with Unequal Currents—Magnetic Field Between[194]

Six pairs of electromagnets are shown below. The current in the left electromagnet is one amp and the current in the right one is two amps in each case. They are also separated by the same distance, and they have the same length and diameter. Carefully observe the orientation of the coil and direction of current flow.

Rank the magnetic fields at the midpoints between the electromagnets from the largest to the smallest. (The direction to the right is positive, and –5 < -3.)

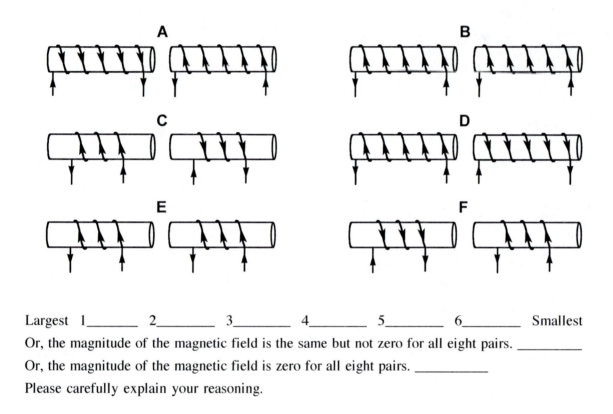

Largest 1_____ 2_____ 3_____ 4_____ 5_____ 6_____ Smallest

Or, the magnitude of the magnetic field is the same but not zero for all eight pairs. _____

Or, the magnitude of the magnetic field is zero for all eight pairs. _____

Please carefully explain your reasoning.

How sure were you of your ranking? (circle one)
Basically Guessed				Sure				Very Sure	
1	2	3	4	5	6	7	8	9	10

Graph of Current vs. Time—Induced Current[195]

We have a long wire with a varying current, I, as indicated in the graph below. Near this wire is a square loop, also shown in the diagram below.

Rank the magnitude of the induced current, i, in the square loop from greatest to least for the different time intervals.

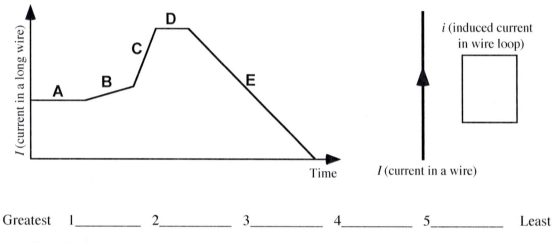

Greatest   1_____   2_____   3_____   4_____   5_____   Least

Or, all of the intervals produce the same induced current. _____

Please carefully explain your reasoning.

How sure were you of your ranking? (circle one)

Basically Guessed   Sure   Very Sure
1   2   3   4   5   6   7   8   9   10

Wave Forms with Same Frequency—Wave Speed[196]

Shown below are six waves, which are all the same kind of wave (e.g., all seismic waves) traveling in various media. The waves all have the same frequency, but their amplitudes, A, and wavelengths, λ, vary as shown in the figures. Specific values for these properties are given in each figure.

Rank these waves from fastest to slowest based on the speed of the waves. That is, put first the wave that is moving fastest and put last the wave that is moving slowest.

A

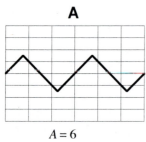

$A = 6$
$\lambda = 3$ m

B

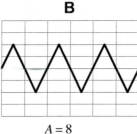

$A = 8$
$\lambda = 2$ m

C

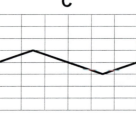

$A = 4$
$\lambda = 6$ m

D

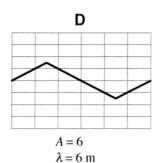

$A = 6$
$\lambda = 6$ m

E

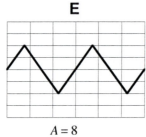

$A = 8$
$\lambda = 3$ m

F

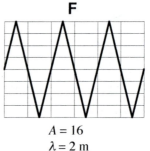

$A = 16$
$\lambda = 2$ m

Fastest 1_____ 2_____ 3_____ 4_____ 5_____ 6_____ Slowest

Or, all of these waves travel at the same speed. ____

Or, all of these waves are at rest. ____

Please carefully explain your reasoning.

How sure were you of your ranking? (circle one)

Basically Guessed				Sure				Very Sure	
1	2	3	4	5	6	7	8	9	10

[196]D. Maloney
Ranking Task Exercises in Physics: Student Edition   196

Wave Pulses—Leading Edge Time to Travel[197]

Shown below are six wave pulses of either triangular or square shape. These pulses, which vary in amplitude, are all sent down identical ropes under equal tension. The ropes are all the same length, and there is no distortion of the pulses as they travel down the ropes.

Rank these pulses, from longest to shortest, on the basis of how long it takes the leading edge to travel 3 m. That is, put first the pulse that takes the most time for the leading edge to travel 3 m, and put last the pulse that travels 3 m in the shortest time.

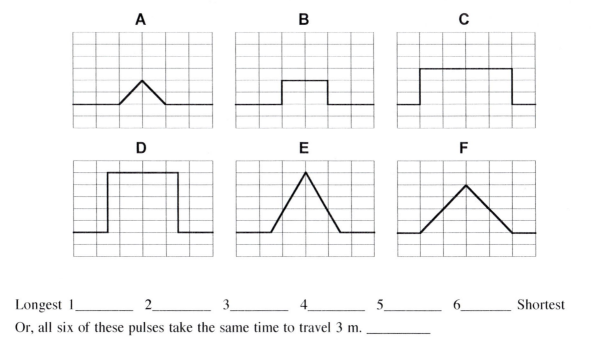

Longest 1_____ 2_____ 3_____ 4_____ 5_____ 6_____ Shortest

Or, all six of these pulses take the same time to travel 3 m. _____

Please carefully explain your reasoning.

How sure were you of your ranking? (circle one)

Basically Guessed   Sure   Very Sure

1   2   3   4   5   6   7   8   9   10

[197] C. Hieggelke, D. Maloney, T. O'Kuma
Ranking Task Exercises in Physics: Student Edition   197

Wave Forms with Same Wavelength—Wave Speed[198]

Shown below are six waves of equal wavelengths traveling in the same medium. Assume the waves are all the same type.

Rank these waves from greatest to lowest according to their speed in the medium.

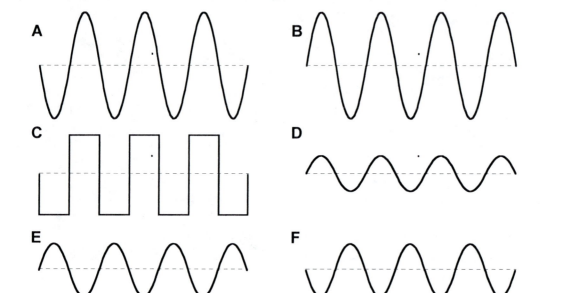

Greatest 1_____ 2_____ 3_____ 4_____ 5_____ 6_____ Lowest

Or, all of the waves have the same wave speed. _____

Please carefully explain your reasoning.

How sure were you of your ranking? (circle one)

Basically Guessed					Sure				Very Sure
1	2	3	4	5	6	7	8	9	10

Wave Forms with Same Wavelength—Wave Energy[199]

Shown are six waves of equal wavelengths traveling in the same medium. Assume the waves are all the same type.

Rank these waves from highest to lowest according to their average wave energy in the medium.

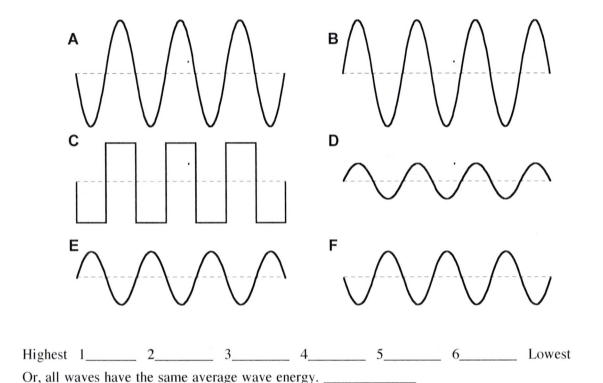

Highest 1_____ 2_____ 3_____ 4_____ 5_____ 6_____ Lowest

Or, all waves have the same average wave energy. _____

Please carefully explain your reasoning.

How sure were you of your ranking? (circle one)

Basically Guessed				Sure				Very Sure	
1	2	3	4	5	6	7	8	9	10

[199] I. Szalai, D. Maloney, T. O'Kuma

Wave Equations—Frequency, Wave Velocity, and Wavelength[200]

Listed below are mathematical representations of one-dimensional wave motion where x and y are in meters, t is in seconds, and the numerical values have appropriate units.

A

$y = 6 \sin (3x-8t)$

B

$y = 6 \sin (8x-3t)$

C

$y = 3 \sin (6x+8t)$

D

$y = 6 \cos (3x+4t)$

E

$y = 8 \sin (6x+8t)$

F

$y = 2 \sin (-3x-4t)$

Rank these mathematical representations of waves, from greatest to least, on the basis of the frequency or how many cycles the waves complete in one second.

Greatest   1_____   2_____   3_____   4_____   5_____   6_____   Least

Or, it is not possible to compare these representations without knowing more. _____

Please carefully explain your reasoning.

Rank these mathematical representations of waves, from greatest to least, on the basis of the wave velocity.

Greatest   1_____   2_____   3_____   4_____   5_____   6_____   Least

Or, it is not possible to compare these representations without knowing more. _____

Please carefully explain your reasoning.

Rank these mathematical representations of waves, from greatest to least, on the basis of the wavelengths of the waves.

Greatest   1_____   2_____   3_____   4_____   5_____   6_____   Least

Or, it is not possible to compare these representations without knowing more. _____

Please carefully explain your reasoning.

How sure were you of your ranking? (circle one)

Basically Guessed   Sure   Very Sure

1   2   3   4   5   6   7   8   9   10

Pairs of Transverse Waves—Superposition[201]

Shown below are six different pairs of rectangular transverse wave pulses that vary in height (*H*) and length (*L*). Specific values for the heights and lengths are given in each figure for each pulse. In each pair the pulses are moving toward each other. At some point in time the pulses meet and interact (interfere) with each other.

Rank these pairs, from highest to lowest, on the basis of the maximum (peak) height of the combined (resultant) pulse at the center of the combined waves at the time when the centers of the pairs coincide.

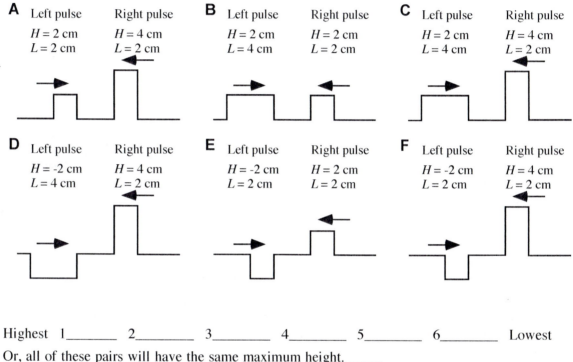

Highest 1_____ 2_____ 3_____ 4_____ 5_____ 6_____ Lowest

Or, all of these pairs will have the same maximum height._____

Please carefully explain your reasoning.

How sure were you of your ranking? (circle one)

Basically Guessed				Sure				Very Sure	
1	2	3	4	5	6	7	8	9	10

[201] C. Hieggelke, D. Maloney, T. O'Kuma
Ranking Task Exercises in Physics: Student Edition   201

Two In-Phase Wave Sources—Intensity at a Point I[202]

Shown below is a situation where two point sources are generating waves of the same frequency and amplitude. The two sources are in phase with each other, and are sitting side-by-side on a line. Letters identify six other points in this region. The distances between the various points can be determined from the grid shown. The wavelength of the waves generated by the sources is equal to the distance between the two sources.

Rank these situations, from greatest to least, on the basis of the intensity at the points identified.

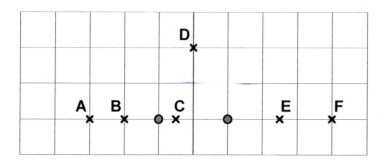

Greatest 1_____ 2_____ 3_____ 4_____ 5_____ 6_____ Least

Or, all six locations have the same intensity._____

Or, the intensity is zero at all six locations._____

Please carefully explain your reasoning.

How sure were you of your ranking? (circle one)

Basically Guessed					Sure				Very Sure	
1	2	3	4	5	6	7	8	9	10

Two Out of Phase Wave Sources—Intensity at a Point[203]

Shown below is a situation where two point sources are generating waves of the same frequency and amplitude. The two sources are 180° out of phase with each other, and are sitting side-by-side on a line. Letters identify six other points in the region. The distances between the various points can be determined from the grid shown. The wavelength of the waves generated by the sources is equal to the distance between the two sources.

Rank these situations, from greatest to least, on the basis of the intensity at the points identified.

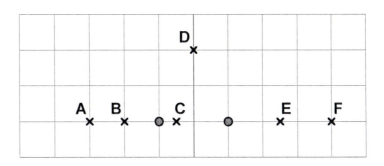

Greatest 1_____ 2_____ 3_____ 4_____ 5_____ 6_____ Least

Or, all six locations have the same intensity._____

Or, the intensity is zero at all six locations._____

Please carefully explain your reasoning.

How sure were you of your ranking? (circle one)
Basically Guessed   Sure   Very Sure
1   2   3   4   5   6   7   8   9   10

---

[203] D. Maloney, C. Hieggelke
Ranking Task Exercises in Physics: Student Edition   203

Two In-Phase Wave Sources—Intensity at a Point II[204]

Shown below is a situation where two point sources are generating waves of the same frequency and amplitude. The two sources are in phase with each other, and are sitting side-by-side on a line. Letters identify six other points in this region. The distances between the various points can be determined from the grid shown. The wavelength of the waves generated by the sources is equal to the distance between the grid lines.

Rank these situations, from greatest to least, on the basis of the intensity at the points identified.

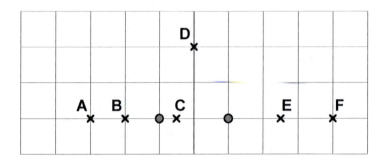

Greatest 1_____ 2_____ 3_____ 4_____ 5_____ 6_____ Least

Or, all six locations have the same intensity._____

Or, the intensity is zero at all six locations._____

Please carefully explain your reasoning.

How sure were you of your ranking? (circle one)
Basically Guessed   Sure   Very Sure
1   2   3   4   5   6   7   8   9   10

[204] D. Maloney, C. Hieggelke

Standing Waves—Frequency[205]

The figures below show systems of standing waves set up in strings, fixed at both ends, under tension. All of the strings are identical except for their lengths and are under the same tension. The variables in these situations, in addition to the lengths (L) of the strings, are the amplitudes (A) at the antinodes and the number of nodes.

Rank these systems, from greatest to least, on the basis of the frequencies of the waves.

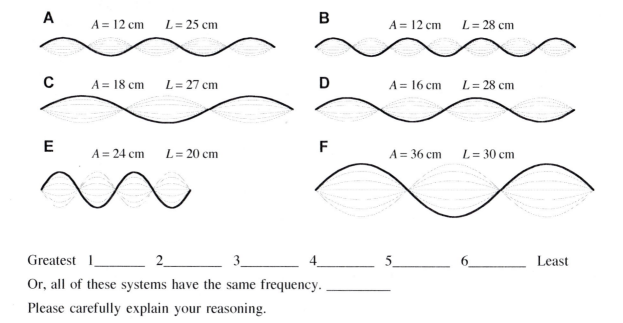

A $A = 12$ cm $L = 25$ cm

B $A = 12$ cm $L = 28$ cm

C $A = 18$ cm $L = 27$ cm

D $A = 16$ cm $L = 28$ cm

E $A = 24$ cm $L = 20$ cm

F $A = 36$ cm $L = 30$ cm

Greatest 1_____ 2_____ 3_____ 4_____ 5_____ 6_____ Least

Or, all of these systems have the same frequency. _____

Please carefully explain your reasoning.

[205] C. Hieggelke, D. Maloney, T. O'Kuma
Ranking Task Exercises in Physics: Student Edition   205

Standing Waves—Wavelength[206]

The figures below show systems of standing waves set up in strings, fixed at both ends, under tension. All of the strings are identical except for their lengths and are under the same tension. The variables in these situations, in addition to the lengths (L) of the strings, are the amplitudes (A) at the antinodes and the number of nodes.

Rank these systems, from greatest to least, on the basis of the wavelengths of the waves.

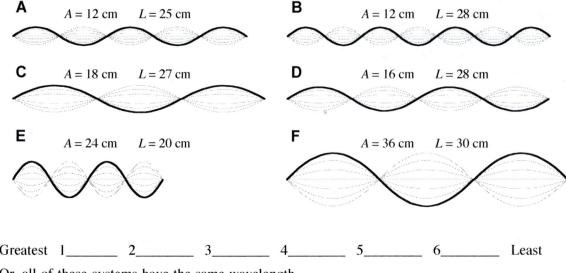

A $A = 12$ cm $L = 25$ cm

B $A = 12$ cm $L = 28$ cm

C $A = 18$ cm $L = 27$ cm

D $A = 16$ cm $L = 28$ cm

E $A = 24$ cm $L = 20$ cm

F $A = 36$ cm $L = 30$ cm

Greatest   1_____   2_____   3_____   4_____   5_____   6_____   Least

Or, all of these systems have the same wavelength. _____

Please carefully explain your reasoning.

How sure were you of your ranking? (circle one)

Basically Guessed   Sure   Very Sure
1   2   3   4   5   6   7   8   9   10

Standing Waves Systems—Wave Speed[207]

Shown below are six standing wave systems in strings. These systems vary in frequency of oscillation, tension in the strings, and number of nodes. The systems are also set up in various strings. The specific values for the string tensions and the frequencies of oscillation are given in each figure. All of the strings have the same length.

Rank these systems, from greatest to least, on the basis of the speeds of the waves in the strings. That is, put first the system whose waves have the greatest speed in their string and put last the system whose waves are traveling slowest in their string.

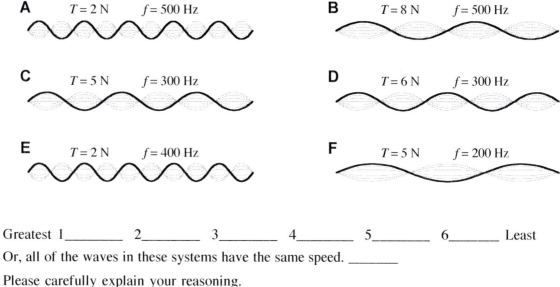

A	$T = 2\,N$	$f = 500\,Hz$		B	$T = 8\,N$	$f = 500\,Hz$
C	$T = 5\,N$	$f = 300\,Hz$		D	$T = 6\,N$	$f = 300\,Hz$
E	$T = 2\,N$	$f = 400\,Hz$		F	$T = 5\,N$	$f = 200\,Hz$

Greatest 1_____ 2_____ 3_____ 4_____ 5_____ 6_____ Least

Or, all of the waves in these systems have the same speed. _____

Please carefully explain your reasoning.

How sure were you of your ranking? (circle one)

Basically Guessed   Sure   Very Sure

1   2   3   4   5   6   7   8   9   10

Person in Plane Mirror—Fraction Visible[208]

Shown below are six arrangements where a student is standing at various distances from one of two plane mirrors. The mirrors are either 25 cm or 50 cm tall. The mirrors are set up so that when the student is at the closest point he can just see the top of his head in the top of the mirror. The student stands 1 m, 2 m or 3 m from the mirrors.

Rank these situations, from greatest to least, on the basis of the fraction of his body that the student can see in the mirror.

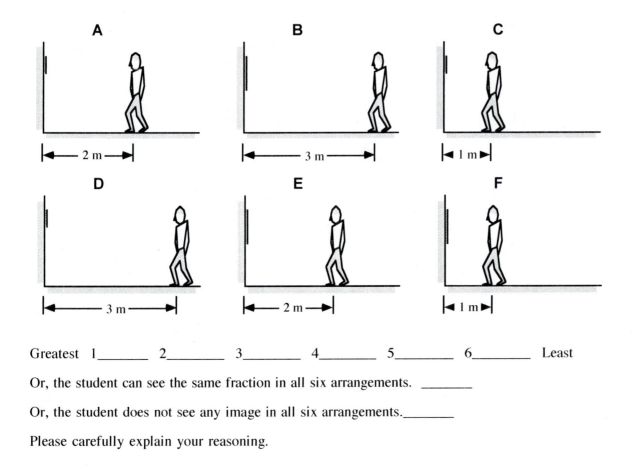

Greatest 1_____ 2_____ 3_____ 4_____ 5_____ 6_____ Least

Or, the student can see the same fraction in all six arrangements. _____

Or, the student does not see any image in all six arrangements._____

Please carefully explain your reasoning.

How sure were you of your ranking? (circle one)
Basically Guessed   Sure   Very Sure
1   2   3   4   5   6   7   8   9   10

Lenses and Mirrors—Focal Length in Water[209]

The figures below show converging (B) and diverging (A) thin lenses as well as converging (D) and diverging (C) mirrors. They all have the same (magnitude) focal length when placed in air (n=1).

Rank these optical systems, from greatest to least, on the basis of their focal lengths when they are placed in water (n=1.33).

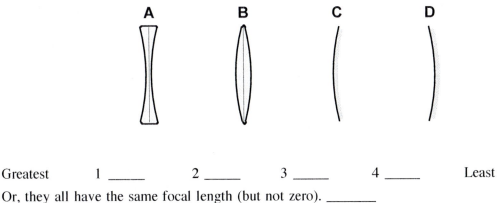

Greatest   1 _____   2 _____   3 _____   4 _____   Least

Or, they all have the same focal length (but not zero). _____

Or, there is no such thing as focal length when these are placed in water. _____

Please carefully explain your reasoning.

How sure were you of your ranking? (circle one)
Basically Guessed   Sure   Very Sure
1   2   3   4   5   6   7   8   9   10

Lenses and Mirrors—Image Distance[210]

The figures below show converging (B) and diverging (A) thin lenses as well as converging (D) and diverging (C) mirrors. They all have the same (magnitude) focal length when placed in air (n=1). An object shown by the vertical arrow in the diagram is placed in front of each optical system at the same distance from the lens or mirror.

Rank these setups, from greatest to least, on the basis of the (magnitude) distance from the image to the center of the lens or mirror.

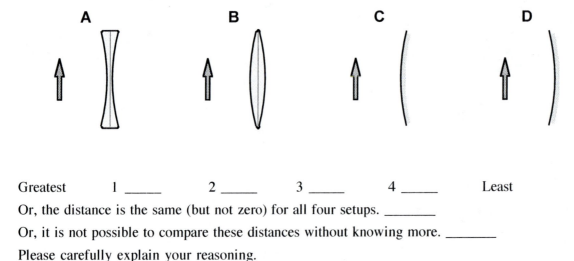

Greatest   1 _____   2 _____   3 _____   4 _____   Least

Or, the distance is the same (but not zero) for all four setups. _____

Or, it is not possible to compare these distances without knowing more. _____

Please carefully explain your reasoning.

How sure were you of your ranking? (circle one)
Basically Guessed   Sure   Very Sure
1   2   3   4   5   6   7   8   9   10

Lenses and Mirrors—Object to Image Distance[211]

The figures below show converging (B) and diverging (A) thin lenses as well as converging (D) and diverging (C) mirrors. They all have the same (magnitude) focal length when placed in air (n=1). An object shown by the vertical arrow in the diagram is placed in front of each optical system at the same distance, greater than the focal length, from the lens or mirror.

Rank these setups, from greatest to least, on the basis of the (magnitude) distance from the object to the image.

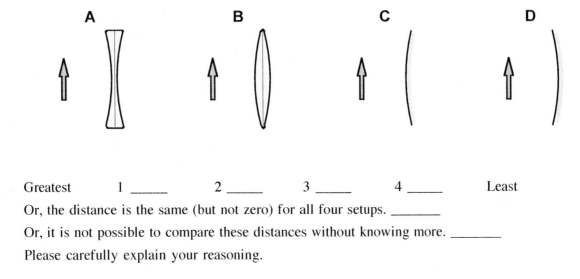

Greatest   1 _____   2 _____   3 _____   4 _____   Least

Or, the distance is the same (but not zero) for all four setups. _____

Or, it is not possible to compare these distances without knowing more. _____

Please carefully explain your reasoning.

How sure were you of your ranking? (circle one)

Basically Guessed   Sure   Very Sure
1   2   3   4   5   6   7   8   9   10

[211] C. Hieggelke, D. Maloney
Ranking Task Exercises in Physics: Student Edition   211

Lenses—Object to Image Distance[212]

The figures below show converging or diverging thin lenses. Their focal lengths (magnitude) in air are x, $2x$, or $x/2$ with focal points indicated in the diagrams. An object (not shown) is placed in front of each optical system at the same distance, $3x$, from the center of the lens.

Rank these setups, from greatest to least, on the basis of the (magnitude) distance from the object to the image.

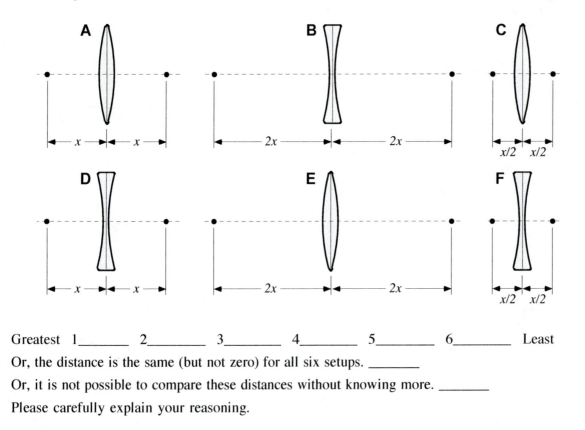

Greatest 1_____ 2_____ 3_____ 4_____ 5_____ 6_____ Least

Or, the distance is the same (but not zero) for all six setups. _____

Or, it is not possible to compare these distances without knowing more. _____

Please carefully explain your reasoning.

How sure were you of your ranking? (circle one)

Basically Guessed				Sure				Very Sure	
1	2	3	4	5	6	7	8	9	10

[212] D. Maloney, C. Hieggelke

Lenses with Baffles—Image Fraction[213]

Shown below are six arrangements where an object is placed at the same distance from the same converging lens. The objects have two heights—short or tall. All of the objects form images on a screen at the same distance to the right of the lens. An opaque baffle covers some part of the lens. The situations differ in whether the baffle is on the object or image side of the lens and in how much of the lens is covered.

Rank these situations, from greatest to least, on the basis of the fraction of the image—compared to the no baffle case—that is on the screen.

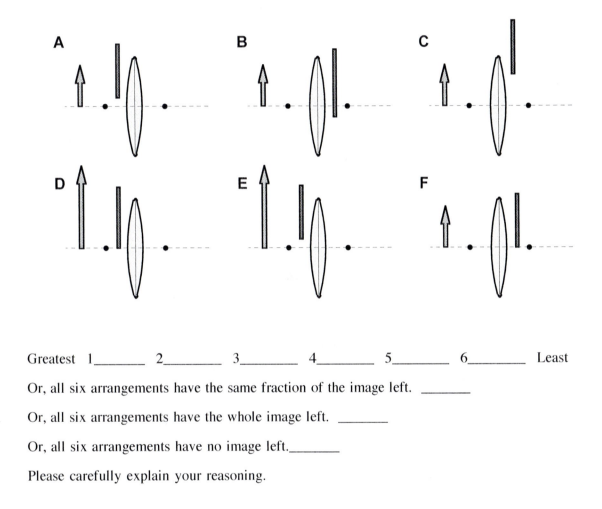

Greatest 1_____ 2_____ 3_____ 4_____ 5_____ 6_____ Least

Or, all six arrangements have the same fraction of the image left. _____

Or, all six arrangements have the whole image left. _____

Or, all six arrangements have no image left._____

Please carefully explain your reasoning.

How sure were you of your ranking? (circle one)

Basically Guessed				Sure				Very Sure	
1	2	3	4	5	6	7	8	9	10

[213] D. Maloney, C. Hieggelke, T O'Kuma

Lens, Object, and Shade—Image Brightness[214]

Parallel light passes by an arrow (the object) through a lens to focus on a screen, as indicated in the figure. In front of the lens, a shade can be lowered covering part or all of the lens. The size and position of the various shades are shown below.

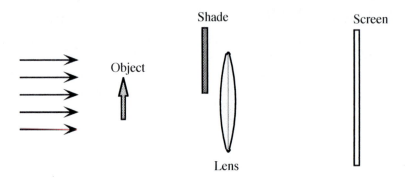

Rank the image from the brightest image to the dimmest image that will be formed on the screen.

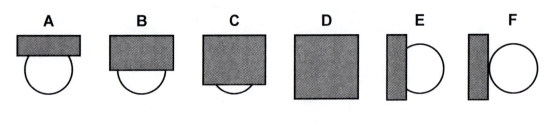

Brightest 1_____ 2_____ 3_____ 4_____ 5_____ 6_____ Dimmest

Or, there is no image formed in any of the cases._____

Or, all images have the same brightness._____

Please carefully explain your reasoning.

How sure were you of your ranking? (circle one)

Basically Guessed				Sure				Very Sure	
1	2	3	4	5	6	7	8	9	10

Cracked or Broken Lenses—Image Size[215]

Parallel light passes by an arrow (the object) through a lens to focus on a screen, as indicated in the figure. All the lenses are identical in focal length. Some lenses are older than others and have parts that are missing or have been cracked over the years as shown below.

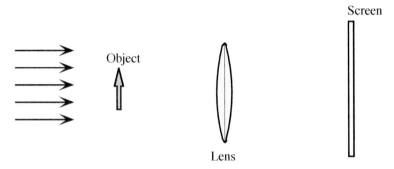

Rank the image from the largest image to the smallest image that will be formed on the screen.

A   B   C   D   E   F

Largest 1_____ 2_____ 3_____ 4_____ 5_____ 6_____ Smallest

Or, there is no image formed by any of these lenses._____

Or, all images are of the same size._____

Please carefully explain your reasoning.

How sure were you of your ranking? (circle one)
Basically Guessed   Sure   Very Sure
1   2   3   4   5   6   7   8   9   10

Cracked or Broken Lenses—Image Brightness[216]

Parallel light passes by an arrow (the object) through a lens to focus on a screen, as indicated in the figure. All the lenses are identical in focal length. Some lenses are older than others and have parts that are missing or have been cracked over the years as shown below.

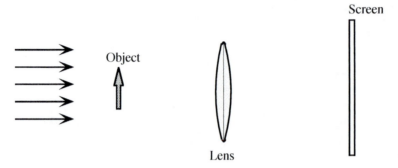

Rank the image from the brightest image to the dimmest image that will be formed on the screen.

A	B	C	D	E	F

Brightest 1_____ 2_____ 3_____ 4_____ 5_____ 6_____ Dimmest

Or, there is no image formed by any of these lenses. _____

Or, all images have the same brightness. _____

Please carefully explain your reasoning.

How sure were you of your ranking? (circle one)

Basically Guessed   Sure   Very Sure
1   2   3   4   5   6   7   8   9   10

Two Lenses—Image Distance to Second Lens[217]

The figures below show pairs of converging or diverging thin lenses. Their focal lengths (magnitude) in air are all f except where indicated in the diagrams as 2f. An object (not shown) is placed to the left of each optical system at the same distance, 2f, from the center of the first lens. Each lens is separated from the other by a distance of 3f/4.

Rank these setups, from greatest to least, on the basis of the (magnitude) distance from the image formed by the first lens to the position of the second lens.

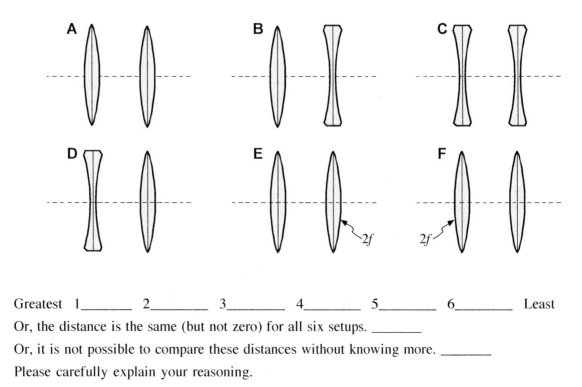

Greatest 1_____ 2_____ 3_____ 4_____ 5_____ 6_____ Least

Or, the distance is the same (but not zero) for all six setups. _____

Or, it is not possible to compare these distances without knowing more. _____

Please carefully explain your reasoning.

How sure were you of your ranking? (circle one)
Basically Guessed				Sure				Very Sure	
1	2	3	4	5	6	7	8	9	10

[217] D. Maloney, C. Hieggelke

Thin Films—Phase Changes[218]

The figures below show light rays striking the surface of a series of transparent thin films with various indices of refraction and thicknesses of t. The light rays reflect and refract as they move through the different films multiple times. Only a few selected rays are shown and the angles are greatly exaggerated to illustrate the particular situations.

Rank these selected rays, from largest to smallest, on the basis of the total number of 180° (or π) phase changes.

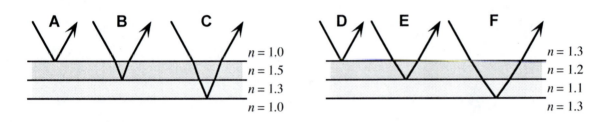

Largest 1_____ 2_____ 3_____ 4_____ 5_____ 6_____ Smallest

Or, they all have the same number of phase changes (but not zero). _____

Or, they all have zero phase changes. _____

Or, it is not possible to compare these phase changes without knowing more. _____

Please carefully explain your reasoning.

How sure were you of your ranking? (circle one)

Basically Guessed   Sure   Very Sure

1   2   3   4   5   6   7   8   9   10

[218] C. Hieggelke, D. Maloney